THE EDGES OF SCIENCE

THE EDGES
OF
SCIENCE

*Crossing the Boundary
from Physics to Metaphysics*

RICHARD MORRIS

PRENTICE
HALL
PRESS

New York · London · Toronto · Sydney · Tokyo · Singapore

Prentice Hall Press
15 Columbus Circle
New York, NY 10023

PRENTICE HALL PRESS and colophons are registered trademarks
of Simon & Schuster, Inc.

Library of Congress Cataloging-in-Publication Data

Morris, Richard, 1939–
The edges of science / Richard Morris.
p.   cm.
Includes bibliographical references.
ISBN 0-13-235029-7
1. Physics.   2. Cosmology.   3. Science.   I. Title.
QC21.2.M67   1990
530—dc20
89-29736
CIP

Designed by Irving Perkins Associates, Inc.

Manufactured in the United States of America

10 9 8 7 6 5 4 3 2 1

First Edition

Acknowledgments

I would like to thank Rolf Sinclair of the National Science Foundation for sending audiotapes of the symposium, "The Edges of Science," held at a meeting of the American Association for the Advancement of Science in 1988.

In addition, I would like to thank the following scientists for sending reprints or preprints of their papers or for discussing their work with me: Hans Dehmelt, Jonathan Dorfan, David N. Schramm, Sidney Coleman, Edwin L. Turner, James B. Hartle, Hyron Spinrad, Jay M. Pasachoff, Bernard Sadoulet, and Blas Cabrera.

CONTENTS

Preface

"The Edges of Science" was the title of a symposium organized by National Science Foundation Physics Program Director Rolf Sinclair and held at a meeting of the American Association for the Advancement of Science in 1988. I was not present at that meeting, but when I listened to an audiotape of the symposium later that year, I was struck by the distinctions Sinclair made between what he called the "edges," the "frontiers," and the "boundaries" of science. His use of this terminology made me suddenly realize that today science is reaching out toward new knowledge, in not just one, but a great variety of different ways. At least this is what seems to be happening in certain scientific disciplines, such as high-energy particle physics, astrophysics, and cosmology.

This has not always been the case. In the past, scientists have, more often than not, been content to seek new knowledge in a relatively straightforward, traditional manner. Experiments would be performed, hypotheses would be tested, and science would inch forward as new data were accumulated. Every now and then, a sudden, great leap would occur, usually when some significant and unexpected new discovery was made.

Today, scientific progress does not always follow the traditional pattern. In fact, there are some scientific fields in which the frontiers have been pushed so far forward that scientists have found themselves asking questions that have always been considered to be metaphysical, not scientific, in nature. At the same time, theoretical speculation has begun to progress so rapidly that often it outpaces experiment. There now exist theories, taken quite seriously by physicists, that are extremely difficult to test experimentally. Other theories are so far advanced that scientists suspect that they may not be capable of yielding predictions that can be experimentally tested until some time in the next century.

This situation has naturally evoked a certain amount of controversy. Nobel prize–winning physicists have been so taken back by some of their colleagues' speculation that they call some of the new theories nonsense, or even compare them to exercises in medieval theology. Other, equally eminent, scientists have taken the opposite point of view and express the conviction that the long-sought "theory of everything," a theory from which all the other laws of physics could be deduced, will soon be discovered.

You might think that the search for this theoretical Holy Grail would lie on the farthest frontiers of physics, but it does not. Some scientists, such as the British theoretical physicist Stephen Hawking, have begun to venture even further, into areas where there are not even any theories to guide them. Hawking and some of his colleagues have, in recent years, begun to speculate about such matters as the possible existence of other universes and the origin of space and time. Meanwhile, other scientists have asked questions about the roles played by life and consciousness in our universe. Still others make queries, not about the physical reality in which we live, but about the characteristics other, presumably unobservable, universes might have.

Clearly, to make any sense of the intense experimental and theoretical activity currently taking place, it is necessary to distinguish between the kinds of scientific work that advance scientific frontiers in a more or less traditional manner, and the kinds of speculation that seek to reach out further, and which, as a result, sometimes seem almost metaphysical.

It is not, however, my intention to set up some arbitrary scheme of classification. I am less concerned with placing different kinds of scientific research in one category or another than I am with discussing the unique nature of some of the scientific activity that takes place nowadays, and in asking what the limits of knowledge really are. Thus I will not introduce any definitions at this point. I would prefer to define my terms as I go along, and to give examples of scientific "frontiers," "edges," and so on.

In any case, the reader is likely to find that what I call "frontiers," "edges," "fringes," and "boundaries" will often blur into one another. When this happens, I will not become unduly concerned, nor, I hope, will the reader. I am more interested in discussing the experimental and theoretical research that is being done, and in describing some of the astonishing new discoveries that have recently been made, than in making fine distinctions. To be sure, there are differences between "frontiers" and "edges." However, pursuing distinctions should not become an end in itself.

Although I have titled this book *The Edges of Science*, it deals almost exclusively with work in the fields of physics and cosmology. I don't apologize for the title, however; I don't think it misleading. It is in these fields of science that the most surprising new discoveries are being made. It is here that we encounter the edges of science.

I

PHYSICS
AND
COSMOLOGY
TODAY

1

The Nature of Matter

Scientists often speak of the fundamental "simplicity" of nature. The great majority of them believe that the universe in which we live is constructed according to simple principles. Though the physical phenomena that scientists observe are often very complex, they invariably assume that the basic laws of nature are not.

It is not immediately obvious that nature really is as simple as scientists like to think; indeed, it would be possible to argue that this idea is a kind of philosophical prejudice. After all, the idea of simplicity is not something that can be proved or disproved; it is a metaphysical postulate. Though it is possible to perform experiments that test the predictions of almost any scientific theory, no one has ever devised an experiment that would tell us whether nature is fundamentally "simple" or "complex." In fact, those notions are not even easy to define. It appears that the postulate of "simplicity" is one that has to be accepted on faith.

Indeed, it would be possible for a skeptic to argue that it is not nature that is simple, but the human mind. Such a skeptic might also maintain that the only reason scientists seek to discover "simple" laws is that they would be unable to grasp principles that were really complex. According to this view, scientific theories would not depict nature as it really is, but would instead be simplified, abstract descriptions of a very complex reality.

Although such an argument could not easily be refuted, one does not often hear such ideas expressed. The reason is an obvious one: science has been too successful. The assumption of simplicity may or may not be an idea that is philosophically suspect, but it is an idea that seems to work. By making the assumption that nature works on simple principles, scientists have been able to gain significant insights into such matters as the origin and evolution of the universe, the nature of the forces that act on objects as small as an electron or as large as a galaxy, and the nature of matter.

In other words, it is possible to justify the postulate of simplicity on pragmatic grounds. The idea that nature is basically simple has led to one scientific success after another. It has also prompted scientists to become skeptical of theories that were later shown to be incorrect. The suspicion that certain ideas were "just too complicated" has frequently advanced scientific understanding.

It is not difficult to find examples of this. It was apparent to Galileo that the Ptolemaic system of astronomy, according to which the sun and the planets followed elaborate orbits around the earth, was just too complicated to be true. Consequently he championed the simpler Copernican system, which put the sun, and not the earth, at the center of the solar system.

We encounter numerous other examples of the rejection of complicated ideas in favor of ones that seemed simpler in looking at scientists' attempts to understand the nature of matter. Over and over again, scientists have tried to understand matter in terms of a small number of constituents. Then, as further discoveries were made, these constituents would become more numerous. Finally, things would reach the point where the feeling would become widespread that "things can't be that complicated," and a new, simpler theory would be developed.

In the time of the classical Greeks, it seemed that matter was not so complex a thing. According to Aristotle, for example, all terrestrial objects were made up of only four elements: earth, air, fire, and water. A fifth element, ether, was (or so Aristotle thought) the constituent of the incorruptible heavenly bodies.

By the middle of the seventeenth century, however, it had become apparent that this simple scheme was not workable. The number of basic substances that could be found on the surface of the earth was much greater than four. If one continued to define an "element" as a substance that could not be broken down into simpler components, then the elements were numerous indeed.

By the end of the nineteenth century, scientists had discovered all of the ninety-two naturally occurring elements. The majority of these, they found, were solids such as iron, silver, nickel, boron, carbon, and sulfur. Some, such as hydrogen, oxygen, nitrogen, chlorine, and neon, were gasses. And, finally, there were two, mercury and bromine, that were liquids at ordinary conditions of temperature and pressure.

Though the discovery of the various chemical elements was a scientific advance, the resulting situation was hardly satisfactory. The idea that there were ninety-two basic kinds of matter, rather than just four, made the world seem unnecessarily complicated. Fortunately, matters became simple again when important new discoveries were made by the

British physicists J. J. Thomson, Ernest Rutherford, and James Chadwick. Thomson's discovery of the electron in 1897 was followed by Rutherford's discovery of the proton in 1919. When Chadwick discovered the neutron in 1932, it appeared that science's understanding of the nature of matter was complete. Atoms consisted of tiny nuclei that were surrounded by orbiting electrons. The nuclei, in turn, were composed of protons and neutrons. The ninety-two elements weren't the basic constituents of matter after all. Instead, there were—or so scientists thought—only three particles.

Hydrogen, for example, was made of one proton and one electron, and was the simplest of the elements. Oxygen, on the other hand, was more complex: The nucleus had eight protons and eight neutrons, and eight electrons circled around it. An atom of uranium was even more complex; its nucleus contained 92 protons and 146 neutrons. Since the positively charged protons and the negatively charged electrons had to be equal in number if the atom was to be electrically neutral, it followed that a uranium atom contained 92 electrons also. Thus there were 330 particles in all; however, every one of them was one of the three basic varieties.

PROLIFERATION OF PARTICLES

Almost at once, it became apparent that this simple scheme was inadequate. In fact, in 1932, the same year that the neutron was discovered, the American physicist Carl Anderson found a new particle, the positron, in cosmic rays. The positron was similar to the electron, except that it carried a positive, rather than a negative, electric charge.

It soon became apparent why positrons had not been discovered before—they did not continue to exist for a very long time once they encountered ordinary matter. As soon as a positron encountered an electron, it and the electron annihilated one another, and gamma rays appeared in their place.

If the positron had been discovered in modern times, physicists would surely have named it the "anti-electron," for the positron is the electron's antiparticle. Today, the prefix *anti* is always part of an antiparticle's name. The positron is the only exception, since it has had this name for so long a time that there has never been any serious attempt to change it.

Scientists know now that, for every particle, there also exists an antiparticle. Thus there are protons and antiprotons, neutrons and antineutrons. And of course, all the particles that we will encounter later have antiparticle partners too: for example, I will later speak of such objects as antineutrinos and antiquarks.

Some antiparticles—the positron is a good example—can continue to exist for long periods of time if they happen to be traveling through space, where the density of matter is low, or if they are confined in devices in physicists' laboratories where they encounter only other antiparticles. However, as soon as a particle and its antiparticle meet, they annihilate one another just as the electron and positron do. This process is described by Einstein's famous equation $E = mc^2$. Here, E is energy, m is mass and c is the speed of light. In the metric units used by scientists, mass may be measured in kilograms, while the speed of light is taken to be 300 million meters per second. In this case, energy will be expressed in joules. A joule is defined to be one watt-second. It is equal to about one four-thousandth of a food calorie.* Although one joule is not a very large quantity, it is obvious that a great deal of energy can be released when matter is annihilated.

* The term *calorie* actually has two different meanings. A food calorie is the so-called large calorie (abbreviated Cal.). It is equal to 1,000 "small calories." The small calorie is defined to be the amount of heat required to raise the temperature of one gram of water one centigrade degree. Yes, life would be simpler if two different terms were used.

After all, c², the speed of light squared, is 90 million million,* an enormous number.

If matter can be converted into energy when a particle and its antiparticle encounter one another, one might suspect that the reverse could take place, that matter could be created out of energy. This is indeed the case. A particle-antiparticle pair can be created in this manner, and the amount of energy required to produce them is, naturally, equal to the amount that is released when a pair is annihilated. Particles and antiparticles, incidentally, are always created in pairs. It is not possible to create an electron, or a positron, or an antineutron, or any other particle, alone. There is much more that could be added about the behavior of particles and antiparticles, but perhaps it would be best to save this for later, and to return to the subject under discussion: scientists' attempts to determine the nature of the basic constituents of matter.

In 1936, just four years after the discovery of the positron, Carl Anderson discovered yet another new particle. This particle resembled the electron, and possessed the same negative charge, but was 207 times as heavy. Originally, the new particle was called the mu meson (later reclassified and renamed muon). *Mu* is one of the letters of the Greek alphabet. (Physicists frequently make use of Greek letters in mathematical equations, which sometimes makes these equations seem more esoteric than they really are, and often use Greek letters to designate subatomic particles.) *Meson* comes from a Greek word meaning "intermediate." This was a reference to the fact that the new particle had a mass much greater than that of the electron, but much less than that of a proton or neutron. Protons and neutrons, by the way, are about equal in mass; they are both approximately 1,800 times as heavy as the electron.

By 1936, then, the number of elementary particles had

* Or, to be more precise, 90 million million meters² per second per second.

already grown from three to five, to include the electron, proton, neutron, positron, and muon. And of course the discovery of the positron suggested that other antiparticles might exist also. In addition, there was yet another particle, whose existence was still hypothetical. In 1930, the Austrian physicist Wolfgang Pauli had pointed out that certain puzzling features of radioactive decays could be explained if one assumed that there existed a particle called the neutrino. However, as it turned out, the neutrino was not to be discovered until 1956.

BARYONS, MESONS, LEPTONS, AND OTHER BEASTS

If the list of fundamental particles had turned out to have no more than five or six entries, or even eight or ten, physicists would most likely have been able to consider them all to be elementary. Unfortunately, as the years passed, the number of known particles increased beyond all reason. By 1960, scores of particles had been discovered. By the early 1970s, the number of "elementary" particles that had been seen by experimenters was in the hundreds.

Some of the particles, known as baryons, seemed to resemble the neutron and the proton, except that they had larger masses. Some of them also had unusual electrical charges. Where the neutron was electrically neutral and the proton carried a positive charge, some of the baryons had negative charges like the much lighter electron, or had twice the positive charge of the proton.

There were also a large number of particles known as mesons. Some of the mesons, such as the pi meson (*pi* is another Greek letter), or pion, were relatively light. The pion had a mass about one-seventh that of the proton. Other mesons, on the other hand, were quite heavy; some of them had masses that were many times greater than those of the proton and neutron.

The particle that Anderson had discovered in 1936 was no

longer grouped with the mesons. Its properties were too unlike theirs. By now scientists realized that it was the electron that the muon most closely resembled. The muon, in fact, could be regarded as a kind of "heavy electron."

A new word, *lepton* (from a Greek word meaning "light"), was invented to describe the electron, the muon, and their associated neutrinos. In 1962, it had been established that neutrinos came in two different varieties, the electron neutrino and the muon neutrino. These two particles, it seemed, were not the same, and they participated in different kinds of reactions.

In 1975, another electronlike particle, the tau particle, or tauon (*tau* is yet another Greek letter) was discovered. As I write this, the tau neutrino has not yet been discovered, however, it is assumed that it must also exist; it would be very surprising if it turned out that electrons and muons had associated neutrinos and the tau did not.

The number of "known" leptons, therefore, currently stands at six: the electron, muon, tau, and three kinds of neutrinos. Naturally, there are six antiparticles also: the positron, the antimuon, the antitau, and three kinds of antineutrinos. However, since particles and antiparticles are so much alike, physicists generally speak of six leptons rather than twelve.

Matter, then, is made of baryons, mesons, and leptons. Although there are only six leptons, each of the other categories has hundreds of members. It sounds too complicated to be believable. At least, no physicists who think that the laws of nature are basically simple could possibly convince themselves that nature had so many fundamental constituents. To make matters worse, many of these supposedly "elementary" particles seem to play no significant role in the scheme of things. If they did not exist, the world around us would seem to be exactly the same—to everyone except experimental physicists, that is.

For example, the muon is a short-lived particle that decays into an electron, a neutrino, and an antineutrino in about one five-hundred-thousandth of a second. If muons did not exist, the properties of ordinary matter would not be altered in the least.

If the existence of so many "elementary" particles made matters complicated, matters were made even worse by the fact that the great majority of the particles decayed into other particles within tiny fractions of a second after they were created. Yet the particles into which they decayed were not simpler constituents of the original particle. This was made obvious by the fact that particles did not always decay in the same way. For example, a pion could decay into an electron and a neutrino, or into a muon and a neutrino, or even into a different kind of pion, accompanied by an electron and a neutrino. Obviously, the original pion could not have been made of all these different things at the same time. Furthermore, there were theoretical reasons for believing that a pion was not a composite of other known particles. There seemed to be no way that an electron or a neutrino could be confined within it.

THE EIGHTFOLD WAY

Scientists realized that, if they were to make progress toward gaining any real understanding of the nature of matter, it would be necessary to bring some order to all this chaos, but it seemed premature to attempt to invent a theory that would explain why so many particles existed. As yet, too little was understood about their behavior. However, the particles could be classified and grouped together in certain natural ways. Each particle, after all, had a set of unique properties. It had a mass. Either it was electrically neutral, or it had a positive or negative charge. Furthermore, each of the elementary particles had a property known as spin. There are certain subtle differences between the spin of an object in the everyday macroscopic world and the spin of subatomic particles. However, the two concepts are similar enough that it is not unreasonable to think of elementary particles as objects that spin on their axes like tiny tops.

Particles had other properties as well. Some of these were given whimsical names such as strangeness; other properties received names that sounded esoteric, but which really weren't, such as isospin. "Strange" particles were ones that decayed much more slowly than physicists expected, while "isospin" is nothing more than a sophisticated way of describing the difference between neutrons and protons, or between other pairs of particles that also seem very similar.

Once the inhabitants of the particle zoo had been placed in different compounds, and their significant characteristics had been labeled, it was possible to take the next step. The animals could be taken out of their compounds again and grouped together in some logical way. The keepers in a real zoo, for example, might notice that lions and leopards seemed to be members of one family, and that there were other characteristics that seemed to make chimpanzees and orangutans seem similar to baboons, monkeys, and gorillas.

Inventing such a classification scheme was such an obvious task that physicists didn't wait very long before accomplishing it. Such a scheme was devised as early as 1961, when the American physicist Murray Gell-Mann and the Israeli physicist Yuval Ne'eman independently discovered that baryons and mesons could be grouped in subfamilies in a particularly natural way. Christened the eightfold way by Gell-Mann, this method quickly proved to be a success. It predicted the existence of hitherto unknown particles, which were soon found by experimenters.

The name "eightfold way" is a pun. Gell-Mann gave the theory that name because it put certain commonly observed mesons and baryons together in groups of eight. He was also aware that the original eightfold way was a program for attaining enlightenment that had been devised by the Buddha around the sixth century B.C. This isn't the last pun that will be encountered in this book, by the way. Puns and other word plays crop up quite frequently in contemporary physics. It is hard to say why this should be the case—perhaps the physicists are trying to convince us that they aren't always the dour fellows that the general public sometimes takes them to be.

Scientists are never satisfied with observing that there are similarities between certain objects, but immediately want to know why these similarities exist. Once it had been established that Gell-Mann's and Ne'eman's eightfold method worked, the next step was to find out *why* it did. In other words, scientists wanted to find out what assumptions they had to make about elementary particles in order to conclude that they would group themselves together in this manner.

In 1964, Gell-Mann and the American physicist George Zweig independently pointed out that the eightfold way could be explained if one assumed that baryons and mesons had constituents that did not resemble any previously known particle. Zweig proposed that these hypothetical constituents be called "aces." Gell-Mann named them quarks. *Quark* is a German word meaning "curds." However, Gell-Mann was not thinking of cottage cheese when he proposed this terminology. He took the term from a passage in James Joyce's novel *Finnegans Wake* concerning the cuckolding of King Mark in the legend of Tristram and Isolde: "Three quarks for Muster Mark."

There were also three quarks in Zweig and Gell-Mann's theory. Called the up, down, and strange quarks, they seemed capable of explaining all the mesons and baryons that were then known to exist. The proton, for example, was made of one down and two up quarks, while the constituents of a positively charged pion (a pion can have a positive or negative charge, or be electrically neutral) were an up and an antidown quark. As one might expect, quarks have their antiparticles too. An antidown is the antiparticle of a down quark.

The names "up" and "down," incidentally, have no particular significance. They are nothing more than arbitrary labels. Instead of "up" and "down," physicists could just as well have called these two particles "one" and "two" or "alpha" and "beta," or even "George" and "Nancy" or "Tristram" and "Isolde." Calling the third quark "strange," on the other hand, does have some significance, since it is one of the constituents of all strange particles. Naturally, the up and strange quarks have antiparticles, just as the down quark does. They are called antiup and antistrange.

At first, many physicists, including Gell-Mann himself, considered the quarks to be nothing more than useful mathematical fictions, not particles that had a real physical existence. In other words, the quark model was thought to be an abstract mathematical scheme, which made some predictions that could be confirmed by experiment, but which had no foundation in reality. As Gell-Mann sometimes put it, baryons and mesons seemed to behave *as if* they had quark constituents.

The reason physicists were so skeptical was that, no matter how hard they looked, they could not confirm the existence of quarks experimentally. Quarks should have been easy to spot because, unlike all other known particles, they were supposed to have fractional charges. The up quark, for example, was supposed to have an electrical charge of $+2/3$ while the down and strange quarks had charges of $-1/3$.

It isn't possible to prove that something *doesn't* exist. For example, there is no way to demonstrate that ghosts don't exist. At best, one can only suggest that it is more plausible to assume that people who report seeing them were probably hallucinating. On the other hand, if one carries out an exhaustive search for something, and fails to find it, it is reasonable to assume that that thing, if it does exist, is exceedingly rare.

Thus when experiment after experiment was performed, and physicists failed to find any free quarks in nature, it seemed reasonable to assume that Gell-Mann and his colleagues were probably right. Quarks were a fiction. After all, the only alternative seemed to be to conclude that they could only exist inside mesons and baryons, never by themselves.

Then, in 1968, an experiment was performed that showed it might be necessary to accept this apparently unreasonable alternative after all. Scientists working at the Stanford Linear Accelerator Center (SLAC) bombarded protons with high-energy electrons, and discovered tiny pointlike charges inside the protons. Quarks were apparently very real after all.

The reason that free quarks are not seen seems to be that the attractive force between quarks is very weak when the quarks are close together, but that it rapidly becomes very strong when the quarks are pulled apart. Thus, if one of the quarks within a proton begins to escape, the other two quarks will pull it back.

Furthermore, it does not seem to be possible to create free quarks by breaking a proton (or some other baryon or meson) apart into its constituents. One can attempt to do this by causing protons to collide with other particles. However, electrons, such as those used in the SLAC experiment, won't work. They simply pass through the protons the way a rifle bullet passes through butter. Nor does causing protons to collide with heavier baryons create free quarks. So much energy is required to break a proton apart that new quarks and antiquarks are created, according to Einstein's formula $E = mc^2$. These newly created quarks then combine with one another to form baryons and mesons. The net result is a number of heavy particles where one existed before.

It is possible to look at this process in another way. Suppose researchers tried to pull two quarks apart. The more they pulled, the stronger the force between them would become. Eventually, so much energy would have been expended that a new quark-antiquark pair could be created. As a result, they would never see any free quarks, only more "ordinary" particles. The new quark-antiquark pair, after all, would stick together just as stubbornly as the pair on which they had been pulling.

A force that drops to zero when two particles are very near one another, and which becomes stronger as their distance increases, behaves in a manner very unlike that of such familiar forces as magnetism or gravity. However, there do exist forces in the everyday world that resemble inter-quark forces. For example, a spring will exert no force whatsoever as long as no one pulls on it. But give the spring a little tug, causing it to expand slightly, and it will begin to pull back. The more the spring is extended, the stronger this force will become.

Naturally, there is always a point at which even the best analogy will break down. If you pull on a spring with enough force, it will eventually break into two pieces. If the spring behaved like a pair of quarks in every respect, this would not happen. Instead, you would find yourself holding a pair of springs, each of them like the original one that you had been attempting to pull apart.

THE CONSTITUENTS OF MATTER

Before long, two additional quarks, named the charm and bottom quarks, were discovered. Just as physicists think that there must be a tau neutrino, they believe that there must be a sixth, top, quark that is paired with the bottom. The top quark is thought to be very massive, which means that a great deal of energy would be required to create it in an experiment. This would explain why it has not yet been seen.

The names "charm," "top," and "bottom" are, of course, just as arbitrary as "up" and "down." Any of these three quarks could easily have been called something else. In fact, for a while there was a movement that insisted on calling two of the new quarks "truth" and "beauty," after Keats's "Ode on a Grecian Urn." However, the more prosaic names "top" and "bottom" won out in the end.

Up, down, strange, charm, bottom, and top are said to

be the six quark "flavors." While it is possible that additional quark flavors will be discovered in the future, there are certain theoretical reasons for believing that the quarks will not proliferate the way that the baryons and mesons did in the 1960s and 1970s. It is believed that, at most, there might be eight or ten quarks, rather than six. And even the possibility that there might be eight is not thought to be very likely. It has not even been suggested, by the way, that there are an odd number of quarks; like leptons, quarks seem to come in pairs.

The basic constituents of matter, then, seem to be twelve in number: six quarks and six leptons.* Except for the forces that act between particles, there is nothing else. And, since muons, tauons, neutrinos, and particles composed of strange, charm, top, and bottom quarks are seen only in the laboratory, one could say that all the things we see in the everyday world have just three constituents: electrons and up and down quarks. Those two quarks, as we have seen, are the constituents of protons and neutrons. With the electron, there are enough particles to make any kind of known atom.

A skeptic might argue that twelve is not so small a number, and add that this number doubles when antiparticles are included. On the other hand, replacing hundreds of subatomic particles with twelve (or twenty-four, counting antiparticles separately) does represent progress. At the very least, a start has been made.

* After this chapter was written, new experimental results were reported which supported the idea that no more than six quarks and six leptons existed. The findings, reported by the Mark II detector team, and by groups of scientists working at the Large Electron-Positron (LEP) collider in Geneva, Switzerland, involved observations of the Z°, a particle we will meet in a subsequent chapter. By examining the manner in which the Z° decayed, the scientists inferred that only three different kinds of neutrino (electron neutrino, muon neutrino, and tauon neutrino) existed. This implied that the leptons were only six in number. And, unless the symmetry between quarks and leptons was somehow broken, there had to be exactly six quarks too.

THE CONSTITUENTS OF MATTER

To review chapter 1:

1. All matter is made up of six quarks and six leptons. Thus there are twelve basic particles (or twenty-four, counting antiparticles separately).
2. The six leptons are the electron, the muon, the tau, and their associated neutrinos.
3. The six flavors of quark are up, down, strange, charm, bottom, and top.

Ordinary matter is made of electrons and up and down quarks. A proton, for example, is composed of one down and two up quarks, the neutron of two downs and one up. If all the basic particles but the electron and the up and down quarks were to disappear overnight, only experimental physicists would know the difference.

2

The Standard Model

I F WE want a complete description of the physical world, and the interactions between particles, it is obviously not sufficient to enumerate the constituents of matter. If we did only that, an important element would be missing. It is also necessary to take the forces that act between particles into account.

There are four known forces: gravity, electromagnetism, and the strong and weak nuclear forces. There does exist some evidence that suggests the possible existence of a fifth force. However, at this writing, this evidence is regarded as controversial. The existence of this force has not been established. However, if it did exist, it would be nothing more than a small correction to the gravitational force. Since it currently plays no role in discussions of the matters treated in this book, I will say no more about it.

Gravity is the weakest of the four forces, but it is the only one that we feel directly. We are also constantly aware of the electromagnetic force, which holds atoms and molecules together, and is responsible for the creation of light, a form of electromagnetic radiation. The strong and weak forces, on the other hand, can only be detected in the laboratory. Although they are intrinsically much stronger than the gravitational and electromagnetic forces, they have extremely short ranges, and their effects are generally felt only at the subnuclear level.

The differences in range are quite dramatic. Gravity can act over distances of millions and even billions of light-years, and hold galaxies and clusters of galaxies together. The strong force, on the other hand, falls off to zero at distances greater than about 10^{-13} centimeters.* The strength of the weak force decreases even more rapidly. This force operates only on a scale less than about 10^{-15} centimeters, a small distance indeed. The diameter of an atomic nucleus is about 10^{-13} centimeters, or approximately a hundred times larger.**

Like the gravitational force, the electromagnetic force is capable of acting over macroscopic distances. Though we are less likely to be immediately aware of it than gravity (unless, of course, we happen to be struck by lightning), its effects permeate our lives. Electricity, obviously, is created by the electromagnetic force. As I mentioned previously, light is a form of electromagnetic radiation. So are infrared and ultraviolet radiation, X rays, gamma rays, and radio waves. It is the electromagnetic force that causes negatively charged electrons to be attracted to positively charged atomic nuclei. It binds atoms together into molecules and

* 10^{13} is the number represented by the digit "1" followed by thirteen zeros: 10,000,000,000,000. 10^{-13} is 1 divided by 10^{13}, or 0.0000000000001.

** Because 10^{-15} is a *smaller* number than 10^{-13}.

also causes molecules to stick to one another. It is the electromagnetic force, in other words, which is responsible for the solidity of solid matter.

The strong force is the force that binds protons and neutrons to one another in atomic nuclei. It acts on baryons and mesons, but not on leptons. The strong force is also the force that binds quarks together within a meson or baryon. In fact, forces between nucleons are viewed as a manifestation of the force between quarks.

Though it is considerably weaker than the strong force, the weak force is equally important—or at least it is important to human beings, since it is reponsible for the nuclear reactions that power our sun. If the weak force did not exist, there might be stars and planets in the universe, but they would be cold, dark bodies.

The characteristics of the forces are summed up in the following table. The units of strength are arbitrary. One could just as easily assign a strength of 1 to the gravitational force. In that case, the strong force would have a strength of 10^{39}. And, of course, "cm" is an abbreviation for centimeters.

THE FOUR FORCES

Force	Strength	Felt by	Range
Strong	1	baryons, mesons, quarks	10^{-13} cm
Electromagnetic	1/100	all charged particles	infinite
Weak	10^{-5}	all particles	10^{-15} cm
Gravity	10^{-39}	all particles	infinite

It might seem odd that gravity should be so important in the universe when the electromagnetic force, which also has an infinite range, is 10^{37} times stronger. The reason for this is simply that matter is electrically neutral. There are as many negatively charged particles in the universe as particles with positive forces. If one variety outnumbered the other, by even a small fraction of 1 percent, electromagnetic forces would operate over large distances and overwhelm the effects of gravity.

ACTION AT A DISTANCE

When Isaac Newton expounded his law of gravity in 1687, he was criticized by certain of his contemporaries, who objected to the idea of "action at a distance." Newton's critics said that if they had some idea as to how gravitational forces could be transmitted, they might be able to take Newton's theory more seriously. The idea of a force that could act across empty space, on the other hand, was simply unacceptable. This, as the German philosopher Gottfried Leibniz put it, made gravity seem like a "perpetual miracle."

Scientists today still have a distaste for the idea of action at a distance. Like Newton's critics, they want to know how a force is transmitted. Fortunately, unlike Newton and his contemporaries, they have a theory that shows how this might be possible. Perhaps one should say "theories," since there are several. These quantum field theories have successfully explained the nature of the forces that act between particles.

The first quantum field theory to be developed was quantum electrodynamics, or QED. QED, which explains the nature of the electromagnetic force, is one of the most successful theories that scientists have ever developed. Its predictions have been experimentally verified to an accuracy of better than one part in a billion, a degree of precision unheard of in other scientific fields.

There also exist theories, modeled after QED, that explain the strong and weak interactions. In fact, as we shall see, there is a quantum field theory that describes the electromagnetic and weak forces within a single framework. Although there does not yet exist any quantum theory of gravity, physicists do not doubt that it will eventually be shown that gravitational forces are transmitted in the same way as the other three. If they are right, and if such a theory is eventually found, then Newton's critics will have at last been answered.

One might think that a theory with a name like "quantum electrodynamics" would be complicated indeed, but this isn't really the case. Like all successful scientific theories, QED is based on concepts that are really quite simple. In fact, there are only two basic assumptions:

1. Forces are transmitted by particles.
2. These particles can pop into existence out of nothing, and then disappear again after the force has been transmitted.

Since the two assumptions are obviously related, we might as well consider the second point first. This is really nothing more than a way of stating Heisenberg's uncertainty principle, which is one of the fundamental postulates of quantum mechanics.

Quantum mechanics is the theory that describes the behavior of all subatomic particles. Heisenberg's uncertainty principle, named after the German physicist Werner Heisenberg, states that it is impossible to determine the position and momentum of a particle at the same time. Equivalently, one could say that it is impossible to determine simultaneously position and velocity. Momentum, after all, is defined to be the product "mass × velocity."

The uncertainty principle has nothing to do with the limitations of scientists' measuring instruments. It states that, even with an apparatus that was perfectly accurate, it would be impossible to know both quantities at the same time. The more exactly velocity (or momentum) is measured, the greater the uncertainty in a particle's position would be. And the more accurately the position was known, the more uncertain the velocity.

When dealing with macroscopic objects, both quantities can be known simultaneously, or at least the uncertainties can be made so small that they are negligible. However, subatomic particles behave differently. If one quantity were known with perfect accuracy, the other could not only not be measured, it couldn't even be defined. If the velocity of an electron were known with absolute precision, nothing could be said about its position; it might be anywhere in the universe.

Although the uncertainty principle is generally stated in terms of position and velocity (or position and momentum), it can also be applied to certain other pairs of quantities. One such pair is time and energy. If we knew the energy of a particle exactly, then we could say nothing about the amount of time that it was likely to remain in that energy state. Conversely, if we knew precisely how long it had been in that state, our ideas about its energy would be fuzzy indeed.

The idea that there is a relationship of this sort between time and energy is no abstract concept. It is something that can actually be observed in the laboratory. For example, it is possible to create pulses of laser light that are very short in duration. When this is done, the pulse is inevitably made up of a bundle of rays of light with different wavelengths and different energies. There is no way that the energy can be determined exactly.

The relationship between time and energy has other important consequences. The uncertainty principle implies that particles can come into existence for short periods of time even when there is not enough energy to create them. In effect, they are created from uncertainties in energy. One could say that they briefly "borrow" the energy required for their creation, and then, a short time later, they pay the "debt" back and disappear again. Since these particles do not have a permanent existence, they are called virtual particles.

Virtual particles are not immune from the principle that particles of matter can only be created in pairs. A virtual electron, or a virtual proton, neutrino, quark is never created alone. It always appears with an antiparticle partner (but we will see later that particles of force can be created alone).

It so happens that there is a visual method for describing particle interactions. This method makes use of Feynman diagrams, which are named after the late American physicist Richard Feynman.

In modern physics, there is no such thing as "nothing." Even in a perfect vacuum, pairs of virtual particles are constantly being created and destroyed. The existence of these particles is no mathematical fiction. Though they cannot be directly observed, the effects that they create are quite real. The assumption that they exist leads to predictions that have been confirmed by experiment to a high degree of accuracy.

The uncertainty principle implies that there is a relation-

(a)

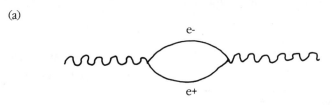

(b)

Two Feynman diagrams. Here, the wavy lines represent gamma rays, e− is the symbol for an electron, and e+ is the symbol for a positron. In (a), two real particles are created from the energy provided by a gamma ray. Some time later, they mutually annihilate each other, and energy is again created in their place. Naturally, this second event does not have to take place; the electron and the positron could have gone off in different directions, never to meet again. In (b), a virtual electron and a virtual positron are created out of borrowed energy. In this case, they must annihilate each other. The uncertainty principle does not provide them with enough time to escape.

ship between the mass of a virtual particle and the length of time that it can exist. Since more energy must be borrowed to create heavy particles than light ones, it follows that the lengths of time that they are allowed to exist are shorter. For example, a virtual electron-positron pair will remain in existence for about 10^{-21} seconds before the two particles disappear again. A virtual proton and a virtual antiproton, on the other hand, will vanish after 10^{-24} seconds (remember that 10^{-24} is the smaller number).

So far, we have considered only particles of matter, such as electrons, but there is no reason why virtual photons, or particles of light, cannot be created, too. It is no contradiction, by the way, to speak of particles of light here, while speaking at other times of light as electromagnetic waves. It was shown early in the twentieth century that light had both a wave and a particle character. In fact, according to quantum mechanics, there is no such thing as a pure wave or a pure particle in the subatomic world. Particles of matter, such as electrons, protons, neutrons, and quarks, also sometimes manifest themselves as waves.

A photon is a particle of light, and light is a manifestation of the electromagnetic force. Thus it would not be inaccurate to say that electromagnetic forces are responsible for the creation of photons. QED goes one step further and says that photons *are* the electromagnetic force.

According to QED and the other quantum field theories, forces are caused by particle exchange. For example, two negatively charged electrons repel one another because virtual photons pass back and forth between them. One electron will emit a virtual photon, and recoil a bit as it does. The photon also gives the second electron a little "kick" as it is absorbed. The two electrons will thus be nudged away from one another.

Note that the creation of virtual photons is a somewhat different process from the creation of particle-antiparticle pairs. Particles of force can be emitted singly; it is not necessary that a particle and an antiparticle be created at the same time.

It is relatively easy to see how an exchange of photons can bring about a repulsion. Attractive forces arise in a similar manner. For example, a negatively charged electron and a positively charged proton also attract one another by exchanging photons. It so happens that there exists an analogy, invented by the British physicist Sir Denys Wilkinson, that might make this a bit easier to visualize. Imagine, Wilkinson says, that two skaters are moving along on a frozen lake. Now suppose that they begin to throw a cricket ball back and forth. It is not hard to see that each skater will recoil a bit when he either throws or catches the ball. The two will gradually be forced apart. But now, Wilkinson says, imagine that the skaters have turned their backs to one another, and throw a boomerang back and forth. One skater throws this object *away* from his partner. The boomerang naturally curves back in the other direction, and is caught by the second skater, who still has his back toward the first. The net result is that there is an attractive force, and the two move closer together.

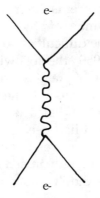

The repulsion between two electrons arises from the exchange of photons. In the diagram above, a photon is emitted by one particle and absorbed by the other. The photon is represented by the wavy line.

It should be emphasized, once again, that no analogy should be pursued too far. In this case, it would be a mistake to assume that photons follow boomeranglike trajectories when they create attractive forces. In fact, the uncertainty principle makes it impossible even to define the trajectory followed by a subatomic particle. However, if we can satisfy ourselves that particle exchange can create attractive forces in the macroscopic world, this should make it easier to accept the idea that particle exchange can produce attraction on the subatomic scale also.

UNIFYING THE FORCES

Quantum electrodynamics is not one of those theories that was worked out by a single individual. Many physicists contributed to its development. In fact, it had quite a checkered history. The basic ideas of QED were worked out during the 1920s and 1930s. However, almost at once physicists encountered difficulties that seemed to make the theory unworkable. As a result, the theory was put back on the shelf, and theoretical physicists turned their attention to other, more tractable problems. Interest in QED revived during the 1940s after a number of physicists, working independently, showed how the difficulties could be avoided.

Thus, by around 1950, the following situation existed: Physicists had a theory of gravity, Einstein's general theory of relativity.* They had a workable theory of the electromagnetic interaction, namely, QED. There was also a theory of the weak interaction that had been proposed by the Italian physicist Enrico Fermi. However, Fermi's theory was capable of describing this process only in a very approximate manner. Finally, physicists didn't understand the strong force

---

* There are two relativity theories. The special theory of relativity deals with the behavior of objects that travel at high velocities; the general theory is a theory of gravitation.

very well at all. To be sure, as long ago as 1935 the Japanese physicist Hideki Yukawa had proposed a theory which showed that the exchange of mesons produced the force between protons and neutrons. However, Yukawa's theory, although it had its successes, didn't seem capable of describing the strong force as accurately as physicists would have liked.

Even if there had been four fully satisfactory theories, one for each of the four forces, there would have been little cause for jubilation. If the laws of nature are basically simple, then it should be possible to find a single theory capable of explaining all the forces. To imagine that gravity, electromagnetism, and the strong and weak forces all operated in different ways would have made the universe seem too complicated.

As we have seen, it was established during the 1960s that baryons and mesons were made of quarks, but this did not immediately do anything to alleviate the unsatisfactory situation with regard to the forces. In fact, it only made matters more difficult since, at first, physicists had no idea what the forces between quarks were like.

The first step toward the unification of the forces was taken in 1967, when the American physicist Steven Weinberg and the Pakistani physicist Abdus Salam independently proposed a combined theory of the electromagnetic and weak forces. Weinberg's and Salam's electroweak theory did describe the weak force more accurately than the theory Fermi had developed. However, it suffered from theoretical problems similar to those that had initially plagued QED. Fortunately, in 1971, the Dutch physicist Gerhard 't Hooft showed how these problems could be eliminated.

According to the theory, the electroweak force (there was now one force, rather than two, in the sense that the weak and electromagnetic forces could now be seen as different aspects of the same interaction) was mediated by a set of four particles. One of these was the familiar photon; the others were designated by the letters W and Z. There were two W particles, one with a positive, and one with a negative, electrical charge. The symbols for these were W^+ and W^-. Since the Z particle was electrically neutral, it was represented by the symbol Z^0.

The electroweak theory turned out to be a resounding success. All three of the new particles were discovered in 1983. Furthermore, all three turned out to be very heavy, about a hundred times more massive than the proton. This was just what physicists had expected; it explained the weak force's short range. It takes a lot of energy to create a massive particle. According to the uncertainty principle, the greater the quantity of energy that must be "borrowed," the shorter the period of time that a virtual particle is allowed to exist. But if the lifespan of a particle is very short, then it will not be able to travel very far before it must again disappear into nothingness. On the other hand, the photon has zero mass. Conse-

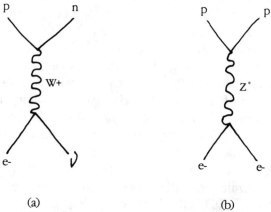

(a)   (b)

Two of the many possible interactions involving W and Z particles are shown. Here v is the Greek letter nu; it represents the neutrino. Note that, since a W particle carries an electric charge, it can change a neutron into a proton or vice versa. In (a), the positively charged proton emits a W+ and becomes a neutron. The negatively charged electron absorbs the charge and is transformed into an electrically neutral neutrino. In (b), no such transformations take place, since the force particle carries no charge.

quently, it may exist for a very long time. It is this which
accounts for the long range of the electromagnetic force.

The relationship between the mass of a particle and the
range of a force becomes a little clearer if we return, once
again, to the analogy of the two skaters. Let us suppose that
the skaters are playing catch with a golf ball. Since the ball is
relatively light, they will be able to throw it quite far, and
interact over moderately great distances. Now suppose that
the skaters decide to throw a medicine ball back and forth
instead. Since the ball is heavy, and cannot be thrown very
far, the skaters must be close together if they are to interact.
If they are too far apart, the second skater will not be able to
catch the ball, and it will go rolling across the ice.

QCD

A theoretical description of the force between quarks was
developed next. During the mid-1970s, theoretical physicists worked out a theory called quantum chromodynamics,
or QCD. According to this theory, quarks come in three
different "colors," which are generally designated red,
green, and blue, by analogy with the three primary colors of
light. The three quark colors obviously have nothing to do
with the colors we see in the everyday world. In fact, it
would be impossible for a quark to have any color at all,

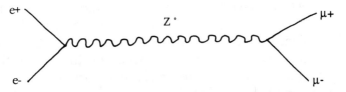

*It is possible to create real, as well as virtual, W and Z particles. In the above
diagram, a positron and an electron annihilate one another, producing, not a
gamma ray, but a Z^0. The Z^0, incidentally, is not required to decay in this
manner; it could have decayed into an electron and a positron again.*

since it is much smaller than the wavelengths of light that correspond to red, green, and blue. The quark colors are nothing more than names for three different kinds of charge that quarks can possess. They could just as well have been called "yes," "no," and "maybe," or "A," "B," and "C," or even "Gertrude," "Alice," and "Virginia."

Obviously, the quark colors, or charges, are not the same as electrical charge, which comes in only two varieties, + and −. However, there is a sense in which the two are analogous. We should not be surprised, however, to discover that quarks can interact in more complicated ways than positive and negative charges. And, indeed, this is what they do. The forces between quarks are not mediated by a single particle, but rather by a set of eight.

The particles that transmit the forces that act between quarks are called gluons. The rationale behind this name should be obvious: gluons "glue" quarks together. Although they come in eight different varieties, they are perfectly analogous to the four particles that mediate the electroweak force.

The interquark color force explains the strong force between baryons and mesons, which can now be understood as a kind of residual force created by the interactions between quarks. A proton and a neutron, or two protons, or two neutrons, will attract one another because there are attractions between the quarks of which they are composed.

The gravitational force is presumably created by exchanges of particles also. Although these particles have never been detected in experiments, and although physicists do not expect to detect them at any time in the foreseeable future, they still have a name. These hypothetical particles are called gravitons. Although there is, as yet, no evidence that they exist, it would be very surprising if it turned out that they did not, for we would not expect one of the four forces to operate in a manner different from that of the other three. If there were no such thing as the graviton, the problem of gravitational action at a distance would be encountered all over again, and the debate that went on between Newton and his contemporaries would have to be reopened.

THE STANDARD MODEL

At this point, it might not be a bad idea to recapitulate as follows:

1. Matter is made of twelve fundamental particles: six quarks and six leptons.
2. There are four forces: strong, weak, electromagnetic, and gravitational. The strong force is really an aspect of the color force that acts between quarks and gluons. Leptons do not experience this force because they have no color. The weak and electromagnetic forces can be described by a single theory. They can be understood as two different manifestations of a single electroweak force.
3. Forces are mediated by the exchange of particles. Twelve force-carrying particles are known: eight gluons, two W particles, the Z^0, and the photon. The graviton, if it exists, would be a thirteenth. For the moment, however, it would be well to leave it out of the list, since there does not yet exist any quantum theory of gravity.

The description of matter and forces that is summed up in the list above has a name. It is called the standard model.

As this chapter is being written, scientists have not discovered any evidence that would contradict any of the theories that make up the standard model. However, many of them do not consider the model to be fully satisfactory. In their view, it has some serious defects.

In the first place, the theories that comprise the standard model do not explain why particles have masses. In fact, in their pure form, they describe only massless particles, and this is of course unrealistic indeed. Some of the particles that we have considered, such as the electron, are not very heavy—but electrons are not massless. Neither are protons and neutrons, and some of the force particles are quite massive. A Z^0, for example, weighs about as much as 100 protons, or 180,000 electrons.

The model can be altered in such a way as to give particles mass. This is done by means of the Higgs mechanism, a theoretical technique named after the British physicist Peter Higgs, who discovered it. The Higgs mechanism involves assuming the existence of an undetected field.* Unlike the electromagnetic, weak, gravitational, and color fields, the Higgs field does not give rise to forces. It "fattens" particles up and provides them with mass instead. However, this method is only partially successful. It does explain why quarks, electrons, muons, taus, and the W and the Z particles might have mass, and leaves the neutrinos massless; but it predicts that gluons should be massless also. Since the color force is short-ranged, we would not expect that this should be the case. As we have seen, it is generally very heavy particles that are associated with forces of short range.

If the Higgs field really exists, then, it must on occasion manifest itself as particles. All quantum fields have this characteristic. However, experiments have not yet turned up any evidence for the Higgs particles' existence.

Perhaps the most serious objection of all is that the Higgs mechanism is introduced in an ad hoc way. There is only one reason for assuming that a Higgs field exists: The standard model does not work without it. One would really like to have better theoretical or experimental justification than this.

There are also other problems with the standard model. For example, it doesn't tell us why quarks and leptons each

---

* When forces propagate through space, physicists often speak of the existence of a field. The nonscientist might suspect the concept of being somewhat tautological. In reality, it is quite useful. An excellent discussion of the role played by the concept of "field" can be found in *The Evolution of Physics* by Albert Einstein and Leo Infeld.

come in six varieties. Here, I am making the assumption that there are only six of each; however, if there were more, the problem would remain. For example, if physicists discovered that there were actually eight leptons and eight quarks, then it would be necessary to explain that. Moreover, the standard model doesn't unify the forces. Ideally, we would like to have one theory that would explain all the forces in nature, rather than three theories.

The standard model does not tell us why some forces should be so strong while others are so weak. This is really just an aspect of the problem of unification, however. If scientists could describe all the forces with a single theory, then that theory would presumably provide an answer to this question.

It is quite obvious that there are problems with the standard model. However, there is nothing particularly unhealthy about such a situation. There are often problems with theories in physics, at least with those that lie on the frontiers of science. If there were no problems, then theoretical speculation and experimental research would come to a halt. Physicists would not know what to do next.

Scientific research consists of solving problems. It is when the problems are most baffling that the research is likely to be most fruitful, and the resulting discoveries the most astonishing. We should be happy that the problems that I have described exist. If they did not, there would be nothing for the present generation of particle physicists to do. Without puzzles to solve, they would not be able to push outward, and extend scientific frontiers.

3

The Big Bang

THE EARTH is constantly bathed by radiation falling on it from every region of the sky. This radiation never varies in intensity. Wherever it is measured, it is found to be equally strong at every hour of the day or night. Nor does it vary according to direction. The radiation that comes to us from the direction of the Big Dipper is no stronger and no weaker than that coming from regions of the sky where we see the constellations Orion or Hydra or, for that matter, from regions where there are no constellations at all. This radiation has another unique characteristic. It is indistinguishable from that which would be emitted by a perfectly black body (a hypothetical object that would reflect no light) at a temperature of 2.7 degrees above absolute zero.

Absolute zero is the lowest possible temperature. It is the temperature at which all molecular motion ceases. It is equal to −273° C (degrees Celsius; this temperature is equal to −460° on the Fahrenheit scale). For convenience, scientists often refer to this temperature as 0 K, where "K" stands for Kelvin (named after the nineteenth-century Scottish physicist Lord Kelvin). The Kelvin and Celsius temperature scales are identical except for the fact that they have different zero points.

Any object with a temperature above absolute zero will emit radiation of some kind. In fact, this is the principle on which the electric light bulb operates. Light is emitted when the filament is heated to a very high temperature. Cold objects also radiate. Naturally, this radiation will not be very intense, and it will not be emitted in the form of visible light. In particular, a body with a temperature of 2.7 Kelvins will emit short radio waves known as microwaves.

Naturally, the intensity of the microwave radiation that falls on the earth is not very great. However, it is measurable, and can be measured quite accurately. Scientists need only point a radio dish toward the sky, and electronically measure the microwaves that fall upon it.

There has never been any significant disagreement among scientists as to the origin of this 2.7 Kelvins cosmic microwave radiation background, which was discovered by the German-American physicist Arno Penzias and the American radio astronomer Robert Wilson in 1964. Only one reasonable explanation for its existence has ever been proposed. It is the afterglow of the big bang explosion in which the universe was born some 10 or 20 billion* years ago. Initially, the universe was in a hot, highly compressed, brightly glowing state. Since then, it has been expanding and cooling off. Today, the universe has cooled to an average temperature of

---

* Here, "billion" is used in the American sense of a thousand million, rather than in the European sense of a million million.

2.7 Kelvins, and what was once an intense burst of radiation has faded to a dim microwave background.

The existence of the background radiation does not constitute the only evidence that a big bang took place. In fact, the first significant discovery that suggested a big bang origin of the universe was made in 1929, some thirty-five years before the microwave background was discovered. In that year, the American astronomer Edwin Hubble discovered that the universe was in a state of rapid expansion, that the galaxies within it were rushing away from one another. Hubble found further that the greater the distance between a galaxy and the earth, the larger the velocity at which it receded.

Naturally, I do not mean to imply that Hubble discovered that our solar system was the center of the universe. The reason that the galaxies seemed to be rushing away from the earth was simply that they were all receding from one another. What Hubble observed was an effect that would have been seen by any astronomer in any galaxy in the universe.

A number of different analogies have been invented to illustrate this point. One might imagine, for example, that a lump of raisin-bread dough is placed in an oven. As the bread rises, the dough expands, and all the raisins recede from one another. If two raisins are initially very close to one another, their velocity of recession will not be very great. If they are nearly touching when the bread begins to rise, they will still be nearly touching when it is taken out of the oven. On the other hand, the distance between raisins on opposite sides of the loaf will increase much more rapidly; the velocity of recession of such a pair of "galaxies" will be greater.

This analogy, too, has its limitations. In particular, one must not be misled into imagining that our universe has boundaries that correspond to the edges of a loaf. In reality, there is no point at which the universe "ends." In fact, the very concept of an "edge" of the universe would be meaningless. If such an edge did exist, what would lie beyond it?

Fortunately, we do not have to deal with the paradox implied by this question. As we shall see when I discuss the implications of Einstein's general theory of relativity, the universe has no boundaries, whether it is finite or infinite.

REDSHIFTS

An obvious conclusion followed from Hubble's discovery that the universe was expanding. If the galaxies were flying away from one another today, then there must have been a time when they were very close together. If one could extrapolate that far back, then there was no reason why one could not look back even further. Presumably there was a time when the galaxies had not yet been created, and when matter existed in a very compressed state. Obviously, if one can calculate when that was, it should be possible to assign a date to the origin of the universe.

Unfortunately, this calculation turns out to be difficult to carry out. Even though more than half a century has passed since Hubble made his great discovery, astronomers still do not agree upon the rate at which the universe is expanding. As a result, there is a great deal of uncertainty as to the amount of time that has passed since the beginning. One set of assumptions can give an age for the universe that is as little as 7 billion years. Another set of assumptions leads to the conclusion that it is 25 billion years old. It appears that the best that one can do is to estimate that the true figure is *probably* somewhere between 10 and 20 billion, but that it could conceivably turn out to be a bit more or a bit less.

The greatest part of the uncertainty arises from problems relating to the measurement of the distances of galaxies. These measurements are exceedingly difficult, and only the distances to the very nearest galaxies are accurately known. It *is* possible to measure velocities of recession with a great deal of precision. However, to determine the expansion rate (and consequently the age of the universe), it would be necessary to know both quantities: speed and distance.

Fortunately, the problem is not nearly as serious as you might think. The question of whether the universe is 10 billion years old, or 15 billion, or 18 billion, isn't really of great importance. Whatever the age of the universe, the dynamics of its expansion are the same. If astronomers eventually discover evidence that allows them to conclude that one figure is likely to be more accurate than the others, then, at worst, they will only have to expand or compress the time scale that they have been using.

Furthermore, even though it may not be possible to say precisely how far away a galaxy is, relative distances can be measured very accurately. It is no problem, for example, to determine that galaxy B is twice as far from the earth as galaxy A.

The reason that this can be done is that, if the expansion of the universe is uniform over large distances—and no one has ever discovered any evidence that it isn't—then distance must be precisely related to a quantity called redshift. In fact, for distances that are not too great, the two quantities are proportional. A doubling of the redshift will imply twice as much distance.

The light from all the galaxies except the very nearest is shifted toward the red. The reason for this is very simple. As we have seen, light consists of electromagnetic waves. These waves have crests and troughs that are analogous to the crests and troughs of ocean waves. The distance between two successive crests or two successive troughs is called the wavelength. When a source of light is stationary, the wavelength remains constant. But suppose that a source of light is moving toward us (or that we are moving toward it; only relative motion matters). This motion will cause successive wave crests to bunch together. The wavelength will be smaller as a result. Suppose, instead, that a source is moving away from us. It is easy to see that, in this case, successive wave crests will be farther apart. As the source emits each successive crest, it will be a little farther away. Thus the wavelength will grow greater.

The longest light waves in the visible spectrum are perceived by us to be red, while the shortest visible wavelengths are violet or blue. Light emitted by a source that is moving rapidly toward us will therefore become bluer, while light that comes from sources that are moving away will redden. Since, with a few exceptions, all the galaxies in the universe are moving away from the earth, their light is shifted toward the red.

This does not imply that light from the most distant galaxies must appear red to the eye (or to a photographic plate) while that from the nearer ones remains more normal. Matters are a little more complicated than that. Very distant galaxies don't look red at all, because as blue light becomes red, radiation from the invisible ultraviolet part of the spectrum becomes blue. The light that comes to us from a distant galaxy will therefore have a full complement of wavelengths, and the galaxy's visual appearance will be very similar to that of a galaxy that is much nearer.

Obviously, it would be impossible to measure redshift by looking at the color of an object. Nevertheless, redshifts can be measured very accurately. Every chemical element emits light of certain specific wavelengths when it is heated. Since the light that comes to us from distant astronomical objects has its source in hot stars, or in glowing clouds of interstellar gas, it is possible to determine, not only the speed of recession of a distant object, but also its chemical composition. We can tell how much helium is present, for example, by looking at the (redshifted) wavelengths that are the characteristic "signature" of the element.

Suppose we observe two automobiles that are traveling at a speed of 50 kilometers per hour. Suppose, further, that they both started out from a city 50 kilometers away. Finally, imagine that one of them has been traveling at the same velocity since its driver started out, but that the other was initially going 80 kilometers per hour, but has since slowed down. Which automobile has been traveling longer? Obviously, the one that has maintained a constant velocity. The one that has slowed down was going faster in the past. Consequently, it has covered the distance in a shorter period of time.

Similarly, calculating the time that has elapsed since the beginning of the universe depends upon two factors: the rate at which it is expanding today, and the degree to which gravity has slowed the expansion. The greater the amount of gravitational "braking" that has taken place, the younger the universe is.

If we know how much matter there is in the universe, it is possible to calculate the effects of this gravitational braking. Again, scientists have been unable to measure the matter density of the universe exactly. As we shall see in future chapters, there seem to be kinds of matter in the universe whose nature they do not understand. So even though there are theoretical reasons for thinking that the universe has a certain specific matter density (these, too, will be discussed in future chapters), yet another uncertainty is introduced into estimates of the universe's age.

However, it is useful to take some figure as an estimate of the time that has elapsed since the big bang. So I will adopt the figure of 15 billion years. It is possible that, during the next few years, this figure might be shown to be in error. Astronomers might find a way of obtaining a much more accurate estimate than they have now. However, if they do, it is not likely that the new figure will be much less or much greater than the one that I have, somewhat arbitrarily, chosen. A figure of 15 billion years for the age of the universe is consistent with observations of the universe's expansion, and with measurements of the age of certain radioactive elements as well; 15 billion years is also a little less than the figure reported for the ages of certain very old stars. However, estimates of those ages have been revised in recent years; they may be younger than astronomers have thought. In any case, I make no claims about the exactness of this figure, and if the universe turns out to be a few billion years older or a few billion years younger, that will not have any significant effect on the discussions that follow.

PRIMORDIAL HELIUM AND DEUTERIUM

Radiation does not have to come from a galaxy to be red-shifted. It is only necessary that it travel through the expanding universe for sufficient periods of time. Thus we would expect that the radio waves that make up the cosmic microwave background should be redshifted too. Microwaves are electromagnetic radiation, and are subject to the same effects as light. Indeed, this is precisely what has happened. The microwave background is really light that was produced in the big bang fireball 15 billion years ago, which has been traveling through space ever since. It is light that was emitted approximately half a million years after the universe began.

Before that time, the universe was filled with free electrons that were moving much too rapidly to be captured by nuclei and form atoms. These electrons interacted with any light that came their way, and absorbed, scattered, and re-emitted it in various different directions. The effect of all this was to produce a kind of cosmic fog. If there had been any conscious observers at this time, they would have found the universe to be nearly opaque, although filled with a brilliant glow. Then, as the universe expanded, its temperature dropped. The same thing happened to the universe that happens to gas that is released from an aerosol can. It, too, cools as it expands, and the can often feels cold as a result. As the universe cooled, the electrons gave up some of their excess energy and began to form atoms. As they did, the fog began to lift. The universe became transparent, and matter and radiation ceased to interact with one another in any significant way.

Observing the microwave background, then, allows scientists to peer back to a time of about half a million years after the beginning, the approximate date at which most of the microwaves that we now observe last interacted with matter. Of course scientists would like to see back even farther if they could—after all, if we can look back this far, why not try and find a way to see even farther?

As it turns out, there are ways that this can be done. Obviously, these don't depend on observations of any kind of radiation. Whatever kind we observed, we could never see very far into the cosmic fog. However, the microscopes that scientists use to examine very small objects don't always make use of light. There are such things as electron microscopes, for example. So perhaps one could attempt to look back in time with some kind of "telescope" that also made use of particles of matter.

The idea is not as crazy as it sounds—in fact, it is quite logical. The methods used to see back to early times don't make use of anything that looks like a telescope. However, they do allow us to draw inferences about the events that were taking place when the universe was very young.

It so happens that there exist substances in the universe that could only have been created in the big bang. Observations that scientists make about the abundance of these substances today allow them to draw inferences about what was taking place when the universe was very young. Specifically, observations of the abundances of the elements helium and lithium, and of deuterium (a form of hydrogen), allow them to "see" back to a time about one minute after the beginning.

There is about one helium nucleus for every ten hydrogen nuclei in the universe. Now, helium nuclei are about four times as heavy as those of hydrogen. Each one is made up of two protons and two neutrons (recall that protons and neutrons are about equal in weight), while an ordinary hydrogen nucleus is nothing other than a single proton. Hydrogen and helium are by far the most abundant elements in the universe. Everything else (including most of the elements that make up earth and its inhabitants) can be thought of as little more than cosmic impurities. The elements other than hydrogen and helium exist in such small quantities that it is accurate to say that the universe is somewhat more than 25 percent helium by weight, and somewhat less than 75 percent hydrogen.

Astronomers have measured the abundance of helium throughout our galaxy, and in other galaxies as well. Helium has been found in old stars, in relatively young ones, in interstellar gas, and in those distant objects known as quasars. Helium nuclei have also been found to be constituents of the cosmic rays that fall on the earth (cosmic "rays" are not really a form of radiation; they consist of rapidly moving particles of numerous different kinds). It doesn't seem to make very much difference where the helium is found. Its relative abundance never seems to vary much. In some places, there may be slightly more of it, in others, slightly less, but the ratio of helium to hydrogen nuclei always remains about the same.

Helium is created in stars. In fact, nuclear reactions that convert hydrogen to helium are responsible for most of the energy that stars produce. However, the amount of helium that could have been produced in this manner can be calculated, and it turns out to be no more than a few percent. The universe has not existed long enough for this figure to be significantly greater. Consequently, if the universe is somewhat more than 25 percent helium now, then it must have been about 25 percent helium at a time near the beginning.

It won't do to assume that the universe was created with a helium content of 25 percent. When the universe was less than one minute old, no helium could have existed. Calculations indicate that, before this time, temperatures were too high and particles of matter were moving around much too rapidly. If a group of neutrons and protons had somehow come together to form a helium nucleus, this nucleus would have collided with other particles almost at once, and would have been blasted apart.

It was only after the one-minute point that helium could exist. By this time, the universe had cooled sufficiently that neutrons and protons could stick together. But the nuclear reactions that led to the formation of helium went on for only a relatively short time. As the universe continued to expand, the average energy of the particles in it dropped still further, and matter became more dispersed. By the time the universe was a few minutes old, helium production had effectively ceased.

The observed abundance of helium, then, provides additional confirmation for the idea that a big bang took place, and allows us to look back to a time when the universe was only a few minutes old as well. But there is another piece of evidence that provides even stronger confirmation for the idea that there was a big bang. This is the existence of deuterium, a variety of hydrogen.

An ordinary hydrogen nucleus consists of a single proton. In deuterium, on the other hand, a proton and a neutron are bound together. Deuterium is a form of hydrogen, and not some other element, because the addition of a neutron to the nucleus does not alter its chemical properties. The nucleus still has a charge of $+1$, and it will still form an atom in which there is a single electron.

Deuterium is not very abundant in our universe. There exists about one deuterium atom for every thirty thousand atoms of ordinary hydrogen. Yet the existence of even tiny quantities of deuterium provides scientists with significant evidence about the big bang. Unlike helium, deuterium cannot be made in stars. The deuterium nucleus is relatively fragile, and it cannot be created, or even exist, in stars. The high temperatures in stellar interiors would cause deuterium nuclei to break apart as soon as they were formed.

The only place that deuterium could have been created is in the big bang.

THE UNIVERSE: OPEN OR CLOSED?

The general theory of relativity, which Einstein propounded in 1915, is an extremely successful and well-confirmed theory of gravitation. During the 1960s and 1970s, a number of different kinds of experiments were carried out to test the theory's predictions. In every case, general relativity passed the test with flying colors.

Like all known theories, general relativity breaks down under certain extreme conditions. As we shall see later, it cannot accurately describe the events that took place very early in the history of the universe, for example, during the first 10^{-43} seconds (this is not an arbitrary number; its significance will be discussed later). Only a theory of quantum gravity could do this. And, as we have seen, such a theory has not yet been developed. However, there is every reason to believe that Einstein's theory gives us an accurate picture of the universe as a whole. Though problems arise when the theory attempts to deal with the very small, or with very early times, general relativity seems to give perfectly accurate results when one is dealing with the very large, including the universe itself.

In particular, Einstein's theory tells us that there are three possible configurations that the universe might have. It is either open, or closed, or flat. The theory does not, however, tell us which of the three possibilities is the case. This is a matter that must be settled empirically. However, general relativity does tell us that the question of whether the universe is open, or closed, or flat, depends upon the quantity of matter that it contains.

A closed universe is one that is finite, but which has no boundaries; it is the three-dimensional analogue of the two-dimensional surface of a sphere. It is useless to try to visualize what the curved space in such a universe would look like. Not even theoretical physicists can do that. However, such a universe can be described mathematically, and its properties can be investigated in detail.

Describing such a universe mathematically is a much less formidable task than one might think. In particular, the concept of curved space is not very abstruse. It means only that the geometry of space is somewhat different from the Euclidian geometry that we learn in high school. For example, there is a theorem in Euclidian geometry that says that the sum of the three angles of a triangle must equal 180°, which is true for any triangle drawn on a flat surface. However, it is not true of a triangle drawn on a curved surface, such as the surface of the earth. In fact, one way of demonstrating that the earth is not flat is to measure the angles between three very distant objects, and then to calculate their sum. Since this sum is greater than 180°, then the earth's surface must be curved.

The geometry of three-dimensional curved space is perfectly analogous. If space is curved, then the angles of a triangle drawn between the centers of three galaxies will never exactly equal 180°. Of course this is an experiment that cannot be carried out in practice. After all, we can measure only one angle. We cannot travel to the other two galaxies to make the same measurements there. And, if they were not *very* far away, the effects would not be great in any case. It is necessary to find some other way to measure the curvature of space in our universe instead.

Before I go on, it would probably be best to digress a bit to make one point perfectly clear. A closed universe is one in which space curves back upon itself, but space is not curved in some fourth spatial dimension. In relativity, as in Newtonian physics, space has only three dimensions. Scientists do speak of four-dimensional spacetime. They do so because the mathematical equations associated with relativity become hopelessly complicated if one attempts to separate the time dimension from the three dimensions of space. However, in reality, the dimensionality of the world of relativity is the same as that of the Newtonian universe.

Though a closed universe is finite, it is *not* true that, if you set out in any particular direction, and traveled long enough, you would eventually return to your starting point from the other direction. A closed universe does not last long enough to be circumnavigated. Not even a ray of light could travel all the way around before the universe collapsed.

A closed universe is one in which the average density of matter exceeds a certain amount. This density has been calculated to be about 5×10^{-27} kilograms per cubic meter (or approximately three hydrogen atoms per cubic yard). If the matter density is greater than this, the average curvature of space will be great enough to close the universe.

The presence of this much matter would have yet another effect. It would create gravitational retarding forces that would eventually cause the expansion of the universe to halt. Since gravity would not cease to act when this happened a phase of contraction would set in. The universe would become smaller and smaller (a closed universe does have a volume, even though it doesn't possess boundaries) until all of the matter that it contained was crushed together in a big crunch.

An open universe is somewhat simpler to describe. Since space does not close in upon itself, such a universe would be infinite in extent. Furthermore, an open universe goes on expanding forever, since the density of matter is not enough to halt the expansion. Gravity may slow down the recession of the galaxies, but it will never stop them completely.

At this point, you might be tempted to ask, "But how can an infinite universe expand?" However, this question is immediately answered as soon as we recall what "expansion" means in this context. An "expanding" universe is one in which the galaxies move away from one another. Obviously this can happen in an open universe as well as in one that is closed. An expanding, infinite universe is simply a universe in which matter becomes progressively more dispersed.

There are a couple of additional points that should probably be emphasized. First, an open universe is *not* one in which some finite number of galaxies expands into a pre-existing void. An open universe, as described by Einstein's theory, is infinite in extent, and it contains an infinite quantity of matter. Of course, to speak of infinite quantities of any kind is to engage in mathematical abstractions. Even if it were determined that the universe seemed to be open, we could never detect galaxies that were an infinite distance away, or be affected by them in any manner.

Another point that should be made is that, whether the universe is open or closed, the big bang was *not* an explosion that hurled matter into pre-existing space. On the contrary, the big bang was an event that took place *everywhere*. This, incidentally, is the reason why the microwave radiation from the big bang fireball falls on the earth from all directions. The region of space in which the big bang took place is not in some specific location billions of light years away; on the contrary, it is all around us.

A flat universe is the simplest one to describe. It would be a universe in which the density of matter is exactly equal to the critical density. A flat universe, in other words, totters on the borderline between open and closed. In a flat universe, the average curvature of space is zero, and the geometry is Euclidian. The sum of the angles of a triangle will equal 180°, or at least they will if the triangle is big enough so that local variations in spatial curvature average out to nothing.

Like the open universe, a flat universe would be infinite. It differs from an open universe in that the expansion, though it never stops, would eventually slow down to such an extent that it would be indistinguishable from zero.

This sounds like a fine distinction, so perhaps it might not be a bad idea to illustrate this point with an example. Let us imagine that an astronomer is observing the recession of the galaxies at some time thousands of billions of years in the future. Now, it so happens that there are good reasons for thinking that neither galaxies nor conscious life will still exist at this point. However, since this is a fantasy, we can imagine anything we want.

If this observer in the far-distant future lives in an open universe, he will always be able to tell that an expansion is taking place. The recession of the galaxies will have slowed, but this effect will still be discernible.* On the other hand, if our astronomer lives in a flat universe, he may not be able to determine whether there is any expansion or not. In a flat universe, the expansion rate never actually reaches zero, but it gets smaller, and smaller, and smaller, with passing time. Eventually it is so small that not even the most accurate instruments would be able to measure it.

A Problem with the Big Bang Theory

The universe in which we live has a very striking feature. It is very nearly flat. Observations show that the density of matter in the universe is almost certainly more than about one-tenth of the critical amount, and almost certainly less than ten times that figure.

Stars provide about 2 percent of the critical density, and there is indirect evidence that mass not incorporated into stars exists in considerable quantities (I will review the evidence for the existence of this mass later). So one-tenth seems to be a reasonable lower limit. Similarly, the actual matter density cannot be much greater than ten times as much. If the universe contained that much matter, we could certainly detect its presence.

Under most circumstances, the discovery that a quantity was equal to some critical value plus or minus a factor of ten would not be considered to be a particularly striking coincidence. For example, if a football team has been averaging

---

* I realize that complications arise if the universe is open, but only just slightly open (that is, very close to being flat). However, I will ignore that possibility so as not to introduce too many fine distinctions in the argument.

twenty points per game, we would not exactly be astonished to discover that the points scored in the last one were between two and two hundred.

However, in the case of the universe, the closeness of the observed matter density to the critical value is striking. The reason is that the ratio between the real density and the critical density changes as the universe evolves. If the discrepancy is less than one part in ten now, then it must have been smaller than one part in 10^{15} when the universe was one second old.

In an open universe the ratio between the actual density and the critical density becomes smaller and smaller as time goes on. As the universe expands, and matter becomes more dispersed, gravitational attractions between galaxies become weaker. The matter that is present in the universe exerts a retarding effect that becomes less and less strong. If the universe originally had, say, 95 percent of the critical density, the ratio will quickly fall to 50 percent, then 25 percent, then 10 percent, and so on.

The opposite effect takes place in a closed universe, where gravity exerts a braking effect that is greater than that which is actually needed. The ratio between the actual density and the critical density becomes larger and larger. This happens even though the universe is expanding because the critical density is not constant; it is a quantity that is related to the expansion rate.

It appears, therefore, that we exist in a very improbable kind of universe, one that was fine-tuned to an accuracy of one part in 10^{15} at a time of one second after the big bang. In fact, this fine-tuning was even greater at earlier times. At some point, when the universe was only a fraction of a second old, it would have been not one part in 10^{15}, but one part in 10^{50}.

If this fine-tuning had not taken place, we would not exist. In a universe that had slightly less matter than ours, the stars and galaxies would never have formed. Matter would have expanded outward at such a rate that gravity could never have created the condensations of hydrogen and helium gas from which the galaxies were formed. On the other hand, if the matter density had differed from the critical value by slightly more than a factor of one part in 10^{15} in the other direction, gravity would have been too strong. The expansion would have halted, and the universe would have collapsed in a big crunch long before life had a chance to evolve.

Even if this were not the case, even if life could exist in a different kind of universe, this accuracy of one part in 10^{15} would still be something that had to be explained. It would not do to call it a coincidence, and leave it at that. Scientists distrust coincidences. When they find that a number is that close to a critical value, they are generally unwilling to believe that this could have happened by chance. They are not satisfied until they find a reason for why the fine-tuning should be that exact.

However, the big bang theory gives no explanation for this degree of accuracy. It says nothing about the rate at which the expansion should have taken place. This is clearly a defect. Even though the theory has not yielded predictions that have been contradicted by experiment, this is a significant fact that it has failed to explain.

This is so conspicuous a failure that it even has a name. The inability of the big bang theory to predict that the matter density of the universe should be so close to the critical value is called the flatness problem. The name is a reference to the fact that a universe with a density this close to the critical value is very nearly flat. A possible solution to this problem will be described in the next chapter, after some other problems with the big bang theory are explored.

4

The Inflationary Universe

THERE IS another major flaw in the big bang theory, known as the horizon problem, which has to do with the fact that the universe looks pretty much the same in every direction. Whichever way we look in the sky, we see approximately the same number of galaxies. To be sure, galaxies are often grouped together in clusters, and there are large regions—enormous "holes" in space—where few or no galaxies are found. However, the farther out scientists look, the more uniform the distribution seems to be. One can compare the appearance of the universe to that of sand on a beach. To an ant, individual grains of sand might appear to be boulders, but to a human being who can look out over distances of hundreds of meters, the beach will seem to be a flat, uniform expanse.

The uniformity of the universe appears even more striking when we examine the microwave background radiation, which originated in a much earlier era than that of galaxy formation. Whatever the direction in which astronomers look, this radiation is approximately the same; its intensity does not vary by more than one part in ten thousand.

In order to see why this uniformity should present a problem, it is necessary to understand the significance of horizons in the universe. Such horizons are not analogous to terrestrial horizons, which result from the curvature of the surface of the earth. They aren't related to curvature at all; on the contrary, they exist because the universe has only existed for a finite period of time.

Let us assume that the universe is approximately 15 billion years old. If it is, then, no matter how powerful the telescopes we construct, we will never be able to see more than 15 billion light-years into space. This follows from the definition of the light-year, the distance that a ray of light will travel in one year.* There may be regions of the universe that are, say, 20 billion light-years away. However, it is not possible for us to see them. It would take 20 billion years for their light to reach us.

On the other hand, if we look in opposite directions, we can see regions of the universe that are 20 or 24 or even 30 billion light-years apart. All we have to do is to look 10 or 12 or 15 billion light-years one way, and then 10 or 12 or 15 billion light-years in the other. This, by the way, is not particularly difficult. Every time astronomers observe the microwave background, they are looking at something that was emitted 15 billion years ago. Meanwhile, galaxies that lie at distances of 12 billion light-years or more have been seen in telescopes.

In other words, we can see regions of the universe that lie

---

* A light-year is equal to about 9.5×10^{12} (9.5 trillion if the American definition of trillion is used) kilometers, or about 6 trillion miles.

beyond one another's horizons. An observer in one of these regions could not see anything in the other. Apparently, these regions can never have been in contact with one another. According to Einstein's special theory of relativity, no signal or causal influence can travel faster than the velocity of light.

If the universe is not 15 billion years old, this does not alter the argument in the least. Only the specific numbers are different. Regions on opposite sides of the sky can never have been in causal contact no matter what the age of the universe. But it is not so easy to explain why these regions should be so similar if one cannot know what the other is doing. What mechanism could be operating that would ensure the uniformity of the microwave radiation to one part in ten thousand? It does no good, by the way, to argue that these regions must have been closer together in the past. Though this is true, horizon distances were shorter also. When the universe was eight and a half years old, for example, the horizon was eight and a half light-years, rather than the 15 billion we observe today.

In addition to the flatness problem and the horizon problem, there exists yet another puzzle which, strictly speaking, is not a problem with the big bang theory itself. This is the fact that, as far as scientists can tell, particles greatly outnumber antiparticles in the universe. Antimatter doesn't seem to exist.

Antimatter would be matter made up of antiparticles. There is no reason why antiparticles should not be able to combine with one another in the same way that particles do, and form atoms and molecules. A positron and an antiproton can theoretically come together and form an atom that would resemble a hydrogen atom in every important respect except one. It would be a positively charged particle that circled a negatively charged nucleus, rather than the reverse. Similarly, it should be possible to assemble something that resembles a helium nucleus from two antiprotons and two antineutrons. With the addition of two orbiting positrons, an atom of antihelium has been created.

If matter and antimatter came into contact, the particles that comprised them would undergo mutual annihilation. The electrons in the matter would annihilate the positrons in the antimatter. Meanwhile protons and antiprotons would undergo mutual annihilation, and neutrons and antineutrons would do the same. As a result, the matter and antimatter would disappear in a burst of energy. An explosion produced in this manner would be many times more powerful than a thermonuclear explosion. When an H-bomb goes off, matter is converted into energy, but the conversion is only partial; a great deal of matter remains.

It appears to be fairly certain that no antimatter exists in our solar system. If it did, it would come into contact with matter from time to time, producing explosions that we would certainly be able to observe. Nor, for that matter, can there be any significant quantities of antimatter in our galaxy. If there were, collisions between clouds of interstellar dust or gas with one another, or with stars, would produce intense bursts of gamma rays, which could easily be detected from the earth.

Conceivably, entire galaxies could be made of antimatter, though this does not appear to be a very likely possibility either. Galaxies collide from time to time, and astronomers have never observed anything that looks like a matter galaxy and an antimatter galaxy coming together.

So the apparent preponderance of matter over antimatter is yet another fact that requires explanation. All of the matter observed today could easily have been created out of energy during the early stages of the big bang. However, when energy is converted into matter, particles and antiparticles are created in equal numbers. If the matter that we observe had such an origin, then where did all the antiparticles go?

GUTs AND THE INFLATIONARY UNIVERSE

Voltaire once observed, of the Holy Roman Empire, that it was "neither holy, nor Roman, nor an empire." Some physicists, such as Stephen Hawking, have noted that the grand unified theories are "not all that grand," nor fully unified either. On the other hand, no one, as far as I know, has denied that they are theories.

The name "grand unified theory," which is often abbreviated to GUT, derives from the fact that these theories represent attempts to unify three of the four forces: the strong force, the weak force, and the electromagnetic force. Obviously, the ideal theory would be one that explained all four forces, including gravity. However, there is nothing wrong with progressing one step at a time, and the electroweak theory was in fact an important step toward unification.

The GUTs—there are several theories that have been proposed—represent an attempt to go beyond the standard model discussed in chapter 2. As yet, no one really knows which, if any, of the GUTs is most likely to be correct. There seem to be certain theoretical problems associated with all of them. In addition GUTs have produced some predictions that do not seem to be confirmed by experiment, though other of their predictions have turned out to be correct.

If the situation regarding the GUTs is somewhat murky, perhaps that is only to be expected. When attempts are made to extend the frontiers of science, problems are always encountered. In any case, in spite of the problems, it seems that the GUTs can't represent an entirely wrong track, for they seem to have implications which solve the cosmological problems I have been describing. Though none of the GUTs can be said to be a really successful theory, they do seem to be capable of explaining why the universe possesses certain observed features.

For one thing, the GUTs seem to explain the preponderance of matter over antimatter. In particular, they tell us that matter and antimatter need not have been created in precisely equal amounts in the big bang. According to the GUTs, it was possible that matter and antimatter could have been created in such a way that there were, say, a billion and one particles of matter for every one of antimatter. Then when the matter and antimatter annihilated each other, the extra particles of matter would have been the only ones left over. Naturally, if such a process did take place, the universe must have contained at least two billion times more particles and antiparticles than it does today, but there is no reason why this could not have been the case.

The GUTs make another prediction that is closely related to this one. If matter and antimatter can be created out of energy in unequal quantities, then it should also be possible for a proton to decay, for example, into a positron and a pion. The assymetry in the creation of matter and antimatter depends upon the existence of a new particle, known as the X particle. If this particle exists, it is not likely to be observed at any time in the foreseeable future. It would have to be very heavy, and the energy required to create it would be greater than that which could be produced in any existing particle accelerator. However, the existence of the X would have observable consequences. In particular, the proton, which physicists have always considered to be perfectly stable, should decay on rare occasions.

Attempts to detect proton decay have been made by a number of different groups of experimental physicists in different countries, but so far none of these experiments has been successful. However, this doesn't necessarily contradict the theoretical prediction. It is possible that the reason that proton decays haven't been seen is simply that they occur too rarely. Different GUTs give different results for the probability of proton decay. So, although the inability of physicists to observe this phenomenon rules out some of the grand unified theories, it doesn't really contradict others.

Nevertheless, the situation is muddled at best. The most significant prediction made by the GUTs has not been confirmed, and it is impossible to tell which of the theories is most likely to be true, if indeed any of them is. In light of these problems, it might seem that it would be just as well to forget about the GUTs for now, and to look for a way to unify all four forces at once. Indeed, as we will see in a subsequent chapter, this is precisely what some theoretical physicists are attempting to do.

THE INFLATIONARY UNIVERSE

Nevertheless, the work that has been done on the GUTs has some important implications. In particular, the GUTs provided the foundation for a theory proposed by MIT physicist Alan Guth in 1980, which showed a way to avoid many of the problems associated with the big bang theory.

Guth discovered that the GUTs seemed to imply that there should have been a very rapid inflationary expansion early during the history of the universe. He found that the quantum fields that permeated the early universe would have created a kind of antigravitational force that would have caused the universe to expand rapidly for a brief period. In particular, the calculations indicated that the inflationary expansion would have set in when the universe was about 10^{-35} seconds old, and would have continued until it had reached an age of approximately 10^{-32} seconds.

According to Guth's theory, the universe could have increased in size by a factor of 10^{50} or more during this brief interval. Then, at the end of this period of inflationary expansion, the driving force exerted by the quantum fields would have faded away, and the universe would have continued to expand at the more leisurely pace that is observed today.

Guth's inflationary universe theory appeared to solve all the problems that I have cited. For example, if the theory was correct, there was no horizon problem. All the regions of the universe that we observe today would have been in contact at times before 10^{-35} seconds, until the inflationary expansion drove them apart. Furthermore, the theory seemed to predict that the average mass density should be very close to, or even equal to, the critical value. In other words, the theory predicted that our universe should be very close to the borderline between an open and a closed universe.

Perhaps the simplest way to understand this latter point is to recall that, if the universe has a mass density that is close to the critical value, then the average curvature of space is very nearly zero. This is precisely what we would expect if an inflationary expansion had taken place, for such an expansion would cause space to flatten out.

Imagine, for example, that a balloon is blown up to a very large size, and that, no matter how much it expands, it never breaks. It is easy to see that, the larger it gets, the flatter its surface will become. Of course, an observer would still be able to tell that it was a balloon. But then, no balloon could expand by a factor of 10^{50} in the way the universe presumably did.

An inflationary universe, then, is one that has had the spatial curvature driven out of it by a rapid expansion. In fact, if the theory is true, then the universe should be so nearly flat that the mass density should be, not one-tenth the critical figure, or ten times that amount, but something much closer to it. In fact, the theory predicts that the density should be precisely the critical value. In practical terms, of course, that only means that it should be very close, since no scientific theory is ever accurate down to the last decimal point.

The theory then seems to solve the horizon and flatness problems. Furthermore, since it is based on the GUTs, the problem of the preponderance of matter over antimatter is solved automatically. We no longer have to wonder why antimatter meteorites never fall on the earth, or why collisions between matter and antimatter galaxies are not observed.

PROBLEMS WITH THE THEORY

You would think that a theory that explained so much would be accepted with delight by scientists. Indeed, the first reactions to Guth's theory were favorable indeed. However, when the theory was investigated in detail, problems began to appear. In fact, it quickly became clear that the inflationary universe theory couldn't possibly be correct.

In particular, the theory predicted that inflationary expansions should have taken place in a lot of separate domains, or spatial bubbles. As these domains expanded, they would have come into contact with one another, and have coalesced into one big universe. Such a process, by the way, is not particularly hard to visualize. We need only imagine expanding soap bubbles that join when they come into contact with one another.

It was obvious that there was something wrong with this picture. The theory predicted that there be domain walls where the bubbles joined. Furthermore, calculations indicated that the individual domains should be much smaller than the observable universe today. The theory said that astronomers should be able to see these domain walls when they looked out into space. And of course they did not.

Fortunately, this problem was soon solved, or at least evaded. Other physicists worked out improved versions of the theory, which avoided this difficulty. The inflationary universe theory was superseded by a new inflationary scenario, which predicted that the individual domains should be much larger than the observable universe, not much smaller. If this was the case, domain walls would most likely not be visible; the chances were that they would lie more than 15 billion light-years away.

I won't discuss the new inflationary scenario in detail, since it has in turn been superseded by other versions of the theory. I will have more to say later about one of these, a theory of chaotic inflation. At this point, however, I think that it would be more fruitful to make some rather general observations about inflationary theories in general than to discuss the individual theories in detail.

PHYSICS OR METAPHYSICS?

There is a sense in which the inflationary universe theories may be unlike most other theories in physics. When I say this I am not referring to the fact that the theory seems to have been invented originally to clear up some difficulties associated with the big bang theory. There is obviously nothing wrong with attempting to find a theory that will explain the observed facts better than the theory we have. The flatness of the universe and its observed uniformity were real physical facts that required explanation.

What I am referring to is the fact that the inflationary theories may not be testable. Ordinarily, a new theory is expected to make predictions that can be tested by experiment. Throughout the history of physics, the best theoretical physicists—and here Einstein is a prime example—have often been very conscientious about suggesting experiments that might confirm or falsify their theories. They felt that theoretical ideas had to be tested if they were to be taken seriously.

This seems not to be the case here. The original inflationary universe theory made just one testable prediction, and that prediction turned out to be false. I am referring to the idea that we should see a lot of individual domains smaller than the observable universe. To be sure, the newer versions of the theory say that these domains should be very large, but this isn't a testable prediction at all. We have no way of knowing whether the nearest domain walls are simply so far away that we can't see them, or whether they don't exist at all.

The inflationary theories do seem to explain certain characteristic features of the observed universe, but this is no test of their validity, since they were invented specifically for this purpose. We can have no way of knowing whether or not there could be a better, completely different, theory that could explain these features just as well.

I am not implying that the inflationary theories should be discarded. There is a great deal about them that is very appealing, and for all their drawbacks, they have been very successful. It is with good reason that they have been incorporated into what has become the standard cosmological outlook.

What I am saying is that inflationary universe theories

have a characteristic encountered with increasing frequency in contemporary physics and cosmology. In recent years, theoretical speculation has had a tendency to outpace experiment. To an increasing extent, new theoretical ideas have had a tendency to gain wide acceptance long before there is any hope of testing them experimentally. In some cases, the members of the scientific community have shown a willingness to accept ideas that cannot be tested at all.

During the course of this book, I will present some other examples of cases where theoretical speculation has reached out so far that experiment has been left far behind, and I will have some more comments to make on the nature of the theoretical endeavor. For the moment, however, it might be best to stick to the topic under discussion. Specifically, I want to ask the question, Are the inflationary universe theories physics, or are they metaphysics?

CREATION OUT OF NOTHING

The metaphysical character of some of the current speculation in the field of cosmology can be seen even more dramatically once we begin to examine some of the speculation to which the acceptance of inflationary ideas has led. In particular, there is a hypothesis, currently very fashionable, which says that the universe may have come into existence out of nothing.

This idea is based on the observation that, if the universe underwent an inflationary expansion at some time in its history, then it could originally have been empty—or very nearly empty—of matter and energy. The universe could have begun as a tiny, expanding bubble of spacetime. All the matter and energy that exist today could have been created during the brief period of inflationary expansion. We could even say that, when the universe went through this phase, matter and energy rushed in to fill the rapidly expanding void.

This is possible because, while the matter content of the universe is positive, gravitational energy makes a negative contribution. Since Einstein's equation $E = mc^2$ implies that matter and energy are only different aspects of the same thing (if we like, we could replace both terms with a single one, perhaps calling it "matter-energy"). Consequently, it should be possible to create huge quantities of matter and energy out of nothing, provided that the positive and negative contributions balance each other. In particular, there is no reason why positive matter and negative gravitational energy cannot be created together.

In order to see why the total energy in the universe should be negative, it is necessary to observe, first, that most of it exists in the form of gravitational energy. The energy in the gravitational fields that hold stars, planets, galaxies, and clusters of galaxies together is far greater than all the other forms of energy combined. This is due to the long range of the gravitational force. Although the gravitational force is relatively weak, every particle in the universe attracts every other. On the other hand, the strong force—to give one example—only operates between protons and neutrons that are practically touching. The electromagnetic force, it is true, is also long-range. However, since matter is electrically neutral, and since magnetic fields in the universe tend to be relatively weak, it does not operate over long distances as the gravitational force does.

In the cosmic scheme of things, then, gravity is far more important than heat, light, chemical energy, or radioactivity. There is much more gravitational energy in the universe than there is nuclear energy. Furthermore, this gravitational energy is negative. It is so large a negative quantity that all the positive contributions of other kinds of energy are unimportant.

The idea of negative energy may seem a little strange at first. However, the concept begins to seem quite reasonable as soon as we ask under what conditions gravitational energy would be zero. The answer to this question is obvious:

when gravitating bodies are so far away from one another that they do not feel any attraction. The gravitational energy in the system comprising the earth and the sun, for example, would be zero if the earth were somehow transported so far away from the sun that it no longer felt any attraction.*

We may observe, next, that if we wanted somehow to move the earth from its present orbit into interstellar space, it would be necessary to expend large quantities of (positive) energy. If we would have to expend energy to get the earth to a position where its energy was zero, it follows that the energy that it possesses now must be negative. It is a simple matter of adding a positive to a negative number. If we add 5 to an unknown quantity, and wind up with nothing, then we must have had -5 to begin with.

The same argument can also be made in reverse. If we now imagine that the earth is initially far out in space, and that it is then allowed to fall back toward the sun, we are led to the same conclusion. If we assume that, aside from some small initial push, no forces act on the earth but gravity, the total energy in the system consisting of the earth and the sun must always remain zero. This follows from the law physicists call the conservation of energy. If we neither add energy to a system, nor allow energy to escape, then the energy in it must always remain the same. Energy may be converted from one form to another, but the total amount is not altered.

However, as the earth falls toward the sun, it will move at greater and greater velocities. It will acquire an energy of motion that constantly increases; but if the total energy in the system remains the same, then the gravitational energy must grow more and more negative.

We must imagine, finally, that the motion of the earth is slowed (perhaps by massive braking rockets, or something

---

* Here I am considering only the energy associated with the attraction of the earth for the sun, not gravitational energy present in the sun or the earth themselves.

of the sort), and that it is allowed to settle into an orbit very much like the one it occupies today. In this case, most of the energy of motion will have been lost, but the large contribution of the gravitational energy will remain.

Now it so happens that we can calculate the contributions of matter and gravitational energy to the matter-energy balance of the universe. It turns out that the contribution of matter is a very large positive number, and that the contribution of gravity is a very large negative quantity. Do they exactly balance one another? No one really knows, but they very well could.

In 1973, the American physicist Edward P. Tryon suggested that the universe might originally have been a quantum fluctuation that grew out of nothing. Tryon's very speculative hypothesis was based on the observation that, according to the Heisenberg uncertainty principle, the less the quantity of energy that is required to create a particle, the longer the time that it is allowed to exist. In particular, if there were such a thing as a particle with zero energy, it would be allowed to exist for an infinite length of time. Obviously, such particles do not exist.* If they did, they would be ghostly entities that could never interact with any kind of matter. On the other hand, a zero-energy universe is a perfectly reasonable idea.

This idea becomes especially plausible when it is examined in the context of the inflationary universe theories. The universe, which perhaps originally contained only a very small number of particles, could have begun as a small quantum fluctuation of some kind. In fact, there exist versions of this hypothesis in which the number of particles is originally two: a particle and its antiparticle.

If the fluctuation persisted long enough for an inflationary

---

* There are particles, such as the photon, that have zero mass. But photons do not have zero energy. In fact, to suggest that they did would be a contradiction; light, after all, is a form of energy, and light is composed of photons.

expansion to begin, the survival of the universe would have been assured. As space expanded, matter and energy could have poured into the universe and filled the rapidly expanding space. Finally, the expansion ceased, and the universe gradually evolved into the cosmos we observe today.

There are many respects in which this scenario seems to be very plausible and appealing. It provides a possible answer to the question, Where did the universe come from? Furthermore, it is quite parsimonious as a hypothesis, for the assumptions on which it is based are few and simple.

On the other hand, it isn't clear whether or not this kind of speculation really has the right to be called "science," or whether it is something more akin to metaphysical philosophy. In scientific fields, theories are supposed to be tested. What experiment could researchers possibly perform to test this one?

Obviously, we can't perform an experiment that requires that we go back in time to see if the universe really did begin this way. Nor can we try and see if this could happen to other universes; we have none on which to experiment. Finally, we don't know whether the matter-energy content of the universe is really zero. When we have two, very large and apparently equal numbers, it may be impossible to tell whether or not they exactly cancel each other. For example, if we have 1 trillion of one quantity, and -1 trillion and 10 of another, we will never be able to tell that they are not equal and opposite if we can only measure them to an accuracy of one part in a billion.

Furthermore, it appears that, under some circumstances, the very concept of the "total energy" of the universe is ambiguous. Einstein's general theory of relativity implies, for example, that, in a closed universe, the concept of "total energy" is meaningless. In the flat universe predicted by the inflationary theories, matters are somewhat simpler, but of course we can't be sure that the universe is precisely flat.

The difference between science and philosophy is supposed to be that scientific ideas are empirically testable, while philosophical ones are not. However, today this principle is being violated, with ever-increasing frequency. It is amusing to note that during the early part of the twentieth century, philosophers were laboring mightily to make their discipline more rigorous. And now, at the end of the same century, the physical scientists that they tried so hard to emulate are introducing untestable ideas into *their* discipline more and more often.

This is not necessarily a lamentable situation—in fact, I would consider it a healthy one. After all, it was a willingness to speculate about the nature of the universe that made the thought of certain past ages great. We would not still read the philosophers of classical Greece today if they had been too timid to speculate. Today, when we have so much more knowledge at our disposal, we should not allow ourselves to be less venturesome than they were. On the other hand, only confusion will result if metaphysics is allowed to masquerade as science. When scientists speculate about ideas that have not been tested, which may not even be testable, they should be willing to admit that the activity in which they are engaging is not quite as "scientific" as some of them would have us think.

II

THE
FRONTIERS
OF
SCIENCE

5

Beyond the Standard Model

During the 1990s, a huge new particle accelerator will be built in Texas at a cost of about $5 billion. It will measure 53 miles in diameter. It will cost about $250 million a year to operate, and when it is running, it will consume more than 30 million watts of power. Called the Superconducting Supercollider, or SSC, it will be used to probe the structure of matter in regions a hundred thousand times smaller than the diameter of a proton. By concentrating large quantities of energy in such tiny volumes, the SSC will reproduce conditions that have not existed since shortly after the creation of the universe. The concentrations of energy that it will produce will equal those that existed in the big bang fireball when the universe was only 10^{-16} seconds old.

The SSC is called "superconducting" because superconducting electromagnets will be used to bend beams of protons into orbits inside two 53-mile rings. The magnets must be superconducting because the power requirements would otherwise be too high (a superconductor is a material through which an electric current can flow without resistance). Once the current is flowing, no additional energy is required to maintain it. For example, if a battery were used to set up a current in a loop of superconducting wire, the current would continue to flow after the battery had been taken away. If a television set could be made of superconducting materials, it would continue to operate after it was unplugged, or at least until the light emanating from the picture tube drained it of energy.

In principle, it would be possible to construct a particle accelerator as large as the SSC with ordinary electromagnets made of copper wire. However, the problems associated with such a design would be enormous. For example, it would require about 4 billion watts of power to operate such an apparatus. In the SSC, on the other hand, most of the electrical power consumed will go to operate the refrigerating devices used to cool the magnets below the critical temperature at which the material in them becomes superconducting.

The SSC will be a supercollider because it will be composed of two rings in which beams of protons will be accelerated in opposite directions. Each ring will be made up of a cryogenic pipeline about 2 feet in diameter, which will surround a much smaller tube that will carry a beam of protons. The protons injected into the SSC will be accelerated around the rings more than 3 million times before they undergo head-on collisions. When such collisions between pairs of protons take place, so much energy will be concentrated in so small a region that, for a brief fraction of a second, energy will be transmitted at a rate greater than the output of all the power plants on earth.

If the SSC had been designed in such a way that a single

beam of protons struck a stationary target, a much smaller quantity of energy—only about 0.5 percent as much— would be released in each collision. It is easy to see why this should be the case. Imagine, for example, that two automobiles collide with one another. If one of them is stationary, then most of the energy of the one that is moving will be expended in pushing the other aside. But if two automobiles collide head on, both will come to a stop, and all their energy of motion will be released.

The SSC will certainly be a marvel of technology, but it will be expensive. Thus we should not be taken aback if we hear a skeptic ask, "Is all this expense really justified? Is it really necessary to spend billions of dollars to send protons crashing into one another? Couldn't research in particle physics be done in some other way?"

Whether the expense is justified is a question that could be argued endlessly. The manner in which we answer it is likely to depend upon the value that we place on knowledge for its own sake. Whatever scientists learn from experiments on the SSC is not likely to have any practical applications for years to come, if indeed it ever does have any. The first particle accelerator, a cyclotron, was built in 1929. In the years since then, scientists have gained a great deal of knowledge about the behavior of fundamental particles. However, technological applications have been practically nonexistent. The development of nuclear bombs and nuclear power, for example, did not depend upon the knowledge gained from experiments with accelerators; nuclear physics and high-energy particle physics are two quite different fields.

In the end, the decision to build, or not to build, expensive pieces of scientific apparatus like the SSC is a political decision, one that is influenced by questions of national prestige as much as it is by purely scientific considerations. Interest in building the SSC increased considerably, for example, when European physicists began to receive Nobel prizes in high-energy physics, a field that had previously been dominated by scientists in the United States. It increased still further when an ambitious program of accelerator construction was undertaken at CERN, the Centre Européen pour la Recherche Nucléaire, near Geneva, and at DESY (Deutsches Electronen Synchroton) in Hamburg. Some suspect that, if American scientists did not have strong competition from Western European scientists, and from the Soviets as well, then the SSC might not have been scheduled for construction until sometime in the next century.

On the other hand, the question of whether the SSC, or something like it, is necessary if further significant advances are to be made in the field of particle physics is somewhat easier to answer—the response must be an emphatic yes. Until the SSC is built, it is not likely that scientists will be able to perform experiments that might lead to discoveries that would advance the frontiers of high-energy particle physics beyond the standard model. In order to probe more deeply into the structure of matter, higher energies are required.

COLLIDING PARTICLES

Physicists have been making particles collide with one another since 1911, when Rutherford used the method to discover the nucleus of the atom. Rutherford directed a beam of alpha particles* at a sheet of gold foil. At the time, particle accelerators had not yet been invented, and the only projectiles available for this kind of experiment were particles emitted in radioactive decays. Rutherford used alpha particles because the only other decay particles known, beta particles (which are nothing but electrons), were too light.

Rutherford found that the energy imparted to the alpha particles by the radioactive substances that emitted them

---

* An alpha particle is composed of two neutrons and two protons; it is identical with the nucleus of a helium atom.

was sufficient to cause the particles to penetrate into the gold atoms that made up the foil. When this happened, some of the particles were deflected by relatively wide angles, while the great majority passed right on through the foil. Rutherford concluded that atoms must contain tiny concentrations of positively charged matter, or nuclei. If the positive charge of an atom was spread out through the atom, as scientists had previously thought, none of the large deflections would have been observed. Finally, Rutherford was able to use the data that he collected on the varying amounts by which different alpha particles were deflected to make detailed calculations about atomic structure. Thus he was able to establish, not only that the atomic nucleus existed, but also to calculate its size.

Since Rutherford's day, scientific apparatus has grown more expensive, and experiments more sophisticated, but the basic experimental pattern has remained the same. For example, the experiment performed at SLAC in 1968, in which it was discovered that pointlike charges (quarks) existed within the proton, operated on exactly the same principle. The only difference was that the SLAC experimenters used a different kind of projectile, which was accelerated to a much higher energy. They worked with electrons, rather than with alpha particles, and accelerated the electrons to high velocities in a 2-mile-long accelerator tube.

PARTICLES AND WAVES

The more deeply we want to probe into matter, the greater the amount of energy required. Although this does not sound particularly unreasonable, it is worth analyzing in some detail. Doing so will throw some light on some of the things that quantum mechanics tells us about the nature of matter.

We might begin by recalling that light has a dual nature. It can be thought of as a stream of particles known as photons, or as a bundle of electromagnetic waves. Until the early years of the twentieth century, scientists would have considered this to be impossible. In their view, things had to be composed of either particles *or* waves; it was impossible for something to be both at the same time. Today we know that their seemingly logical conclusion was wrong. Quantum mechanics tells us that both light and matter are made up of particles and waves at the same time.

Numerous experiments have been performed in which the particle character of light and the wave character of matter are revealed. In some experiments an electron will behave as a particle, for example, when it strikes a fluorescent screen and produces a spot of light. Other experiments can be performed in which beams of electrons clearly show a wavelike character. Furthermore, the electrons that make up the beams can be shown to have characteristics that apply only to waves, such as wavelength and frequency.

In fact, experiments have been performed in which the wave character of single particles have been demonstrated. For example, in 1974, a group of scientists at the Atomic Institute of the Austrian Universities in Vienna found that they could cause a single neutron to pass through a piece of apparatus by two different routes at the same time. Obviously, the neutron could do this only if it was behaving as a wave. Particles do not split up into pieces that then recombine, but a wave can do this easily.

There is a relationship between the wavelength of a particle and its velocity. The faster it is moving, the smaller the wavelength will be. This follows from the fact that faster-moving particles naturally possess more energy of motion. Higher energy always corresponds to shorter wavelengths. This is true of electromagnetic radiation, for example. High-energy ultraviolet rays have wavelengths that are shorter than those of the various colors of visible light. X rays and gamma rays, which have more energy yet, have wavelengths that are even shorter.

This is extremely important because, if researchers want

to "see" an object, they must illuminate it with wavelengths that are shorter than the object itself. This is why viruses cannot be seen with ordinary microscopes; they are smaller than the wavelengths of visible light. In order to make them out, an electron microscope must be used instead. If the electrons are moving rapidly enough, their wavelengths will be short, and clear images can be formed.

In order to "see" the details of the structure of matter on a very small scale, then, particles must be accelerated to high velocities. The only way to accomplish this is to build large, expensive machines. The smaller ones were pushed to their limits long ago.

MEGA, GIGA, AND TERA VOLTS

If a pair of colliding particles possesses large quantities of energy, there is an added benefit: The energy released in the collision can be used to create new particles. Einstein's formula $E = mc^2$ tells us that matter can be created out of energy at any time, but, naturally, if real, not virtual, particles are to be created, sufficient energy must be available.

In the everyday world, energy is measured in such units as kilowatt-hours or calories, but, obviously, it would be silly to speak of a proton as having so many calories of energy. It could be done, but the numbers involved would be cumbersome.

The units of energy used in the field of high-energy particle physics are multiples of the electron volt (abbreviated eV), which is defined to be the amount of energy required to push an electron through a voltage difference of one volt. For example, six electron volts would be expended when one electron passed through a small motor that was connected to a six-volt battery.

Particle accelerators are considerably more powerful than the batteries that one buys at the supermarket. So it should not be surprising that physicists should commonly use large multiples of the electron volt. There are, in fact, three units in common use. The first, a million electron volts, is abbreviated MeV. The symbol for a billion (a thousand million) electron volts is GeV, where the G stands for "giga." At one time, American scientists called this quantity a billion electron volts, and used the symbol BeV, but this only caused confusion because, as I have noted before, the word "billion" has a different meaning in Europe than it does in the United States. Finally, the unit of a million million ("trillion" in the United States, but a "billion" in Europe) electron volts is designated by the symbol TeV, where the T stands, not for "trillion," but for "tera."

All of this can be summed up as follows:

$$1 \text{ MeV} = 1 \text{ million electron volts} = 10^6 \text{ eV};$$
$$1 \text{ GeV} = 1,000 \text{ MeV} = 10^9 \text{ eV};$$
$$1 \text{ TeV} = 1,000 \text{ GeV} = 1,000,000 \text{ MeV} = 10^{12} \text{ eV}.$$

If the energy possessed by a particle can be measured in electron volts, then so can the particle's mass. The equivalence of mass and energy makes this possible. Thus we can say that the electron has a mass of 0.511 MeV (which can also be written as 511,000 eV), while the proton and the neutron have masses of 938 MeV and 940 MeV, respectively.

The masses of some of the fundamental particles are known much more precisely than others. Some have been measured quite accurately: For example, the muon has a mass of 106 MeV (it is about 207 times as heavy as the electron), while the tau weighs in at 1,784 MeV, or 1.784 GeV. The masses of the W and the Z^0 particles have not been measured to the same degree of precision, but it is possible to say that both W particles weigh about 80 GeV, while the Z^0 has a mass of approximately 90 GeV. They are the heaviest fundamental particles known, incidentally. The Z^0, for example, is nearly a hundred times as heavy as the proton.

Since quarks cannot be isolated, it shouldn't be surprising that their masses are known only approximately, since it is only possible to make estimates. The masses of the six quarks are thought to range from about 5 MeV to around 30 GeV. Of course, the reader should not take these figures as gospel. It is perfectly possible that the estimates might change slightly between the time that I write this and the time that this book is published.

Finally, no one really knows what the mass of the neutrino is. All that we can say is that it is either zero, or very small (possibly a few eV). Until a few years ago, it had always been assumed that neutrinos had zero mass, but recent theoretical and experimental work indicates that this might not be the case. As I write this, all that can be said is that, if the neutrino does have a mass, it is so small that no one can accurately measure it.

DISCOVERING NEW PARTICLES

Obviously, powerful accelerators must be built if one is to have any chance of discovering very heavy particles. The massive W and Z^0 particles, for example, were found only after very powerful accelerators were built. And even more powerful ones (like the SSC) are needed if we are to progress further. If a particle has a mass of, say, 200 GeV, one can perform experiments on a 100-GeV accelerator for decades, and it will never be seen. The particle will never appear if the energy required for its creation is not available.

The SSC will cause protons to collide with energies of about 40 TeV. This is approximately ten times greater than the energies produced by the most powerful accelerators at the end of the 1980s. However, we should not jump to the conclusion that the construction of the SSC will make it possible for particles with masses of 40 TeV to be created. Only a fraction of the total energy produced in a proton accelerator can be converted into mass. For example, the Tevatron, an accelerator at the Fermi National Accelerator Laboratory (often called "Fermilab") near Chicago, produces a total energy of about 1.8 TeV, but only about a sixth of this, or approximately 0.3 TeV, is available for particle creation.

The reason for this is simple enough. Protons and anti-protons, after all, are composite particles made up of quarks, antiquarks, and gluons. When they strike one another, all the constituents of one do not collide with all the constituents of the other. On the contrary, the collision will generally take place between only two of them, and a quark may strike an antiquark, for example.

A somewhat ludicrous analogy can be made. Imagine that two people are swinging sacks of volleyballs at each other. When the sacks collide, one of the balls in one sack will generally hit one of the balls in the other. Most of the volley-balls will not participate in the collision at all.

The quantity of energy that is available for the creation of particles depends upon the kind of particle created, as well as upon the power of the accelerator. Theoretical calculations indicate that the SSC should be able to detect Higgs particles (if Higgs particles really exist) with masses up to 1 TeV, previously undiscovered quarks with masses up to 2 TeV and force-carrying particles up to 6 TeV. Thus there is a very good chance that the experiments that will be performed on the SSC will lead to discoveries that will allow scientists to go beyond the standard model.

The High-Energy Frontier

When physicists begin to design the experiments to be performed on the SSC, one of the first priorities will be to attempt to find evidence for the existence of the Higgs particle. According to currently accepted theory, the mass of this particle should be more than 5 GeV but less than 1 TeV. Since the SSC will be capable of producing a 1-TeV Higgs particle, one can reasonably assume that, if the particle exists, it will be seen. Of course there is no guarantee that it will be, and many physicists are skeptical about the particle's existence.

For example, Los Alamos physicist Peter A. Carruthers has described it as something "that people stick into theories just to make the clock work." And University of Michigan physicist Martinus J. G. Veltman has commented, "Indeed, modern theoretical physics is constantly filling the vacuum with so many contraptions such as the Higgs boson that it is amazing a person can even see the stars on a clear night!" (Here, "Higgs boson" is just another way of saying "Higgs particle," and the reference to the vacuum is an allusion to the fact that Higgs fields and Higgs particles are supposed to exist even in a perfect vacuum—that is, in the absence of all other matter.)

Of course, if the Higgs is *not* found, that will constitute an important discovery too. If the particle is not seen, existing theory will have been contradicted, and theoretical physicists will know that they either have to seek some other mechanism that will cause particles to have masses, or modify existing theoretical ideas. Negative evidence of this sort can often be quite important, for it is the realization that existing theory is unsatisfactory that provides the motivation for seeking new theoretical ideas.

A second problem that will be confronted by experiments performed on the SSC will be that of the fundamental constituents of matter. Although most contemporary physicists believe that quarks and leptons are fundamental particles, it is possible that discoveries could be made that would show this assumption to be incorrect. Scientists have thought that they had discovered the basic constituents of matter a number of times in the past. At one time, the atom was supposed to be indivisible. Then neutrons and protons were thought to be fundamental. Today, it is believed that matter is made of quarks and leptons. However, if these particles do have smaller constituents, then the SSC might allow scientists to see them.

Even if no evidence is found for the existence of particles within quarks and leptons, there are still questions to be answered about the constituents of matter. It is not yet possible to be certain precisely how many different kinds of quarks and leptons there are. Why should there be only six of each? Why not eight, or ten, or twenty, or even an infinite number?

Quarks and leptons are often said to come in "families," since they can be grouped in pairs. The electron and the electron neutrino are grouped together, as are the muon and the muon neutrino, and the tau and the tau neutrino. Similarly, the up and down quark are paired with one another, as are the strange and charm quarks. Finally, the bottom quark is paired with the as-yet-undiscovered top.

The existence of three families of leptons and three families of quarks is believed to be something more than mere coincidence. Most physicists see it as evidence of a basic symmetry of nature. Thus, if experiments on the SSC uncovered evidence for either a fourth family of quarks or a fourth family of leptons, physicists would immediately begin looking for evidence for a fourth family of the other. If a fourth family of particles were found, there could presumably be a fifth, a sixth, and so on. Naturally, physicists hope that there are not. Having twelve "fundamental" particles of matter is cumbersome enough. If the number turned out to be very much larger, the problem of the proliferation of particles, such as existed in the days before quarks were discovered, would return again.

As I write this, however, it doesn't seem very likely that this kind of proliferation will take place. Experiments performed at CERN and at DESY seem to hint that a fourth family might exist, though the evidence is indirect. No new particles have been seen, and the arguments for the existence of a fourth family are speculative. Furthermore, there exists a theoretical argument that seems to imply that the number of quark and lepton families can be at most four.*

This theoretical argument provides an example of the way in which the fields of particle physics and cosmology inter-

---

* There is also some new evidence. See the footnote on page 17.

act these days, for it is based on ideas about the expansion of the universe approximately one second after the big bang began.

When the universe was about one second old, its rate of expansion must have depended upon the number of different kinds of neutrinos that existed. The greater the number of different varieties of neutrino, the more rapid the expansion would have been. The rate of expansion, in turn, would have affected the quantities of helium, deuterium, and lithium produced. Thus measurements of the quantities of these substances that exist today provide information about the number of different kinds of neutrinos that exist. If we assume that the symmetry between the number of neutrinos and the number of different families of quarks is maintained, then this gives information about the total number of different kinds of fundamental particles that might exist.

This sounds like a complicated argument, but each step is reasonably straightforward, and unless there is something terribly wrong about scientists' assumptions about conditions in the early universe, it is probably valid. I think that it would therefore be worthwhile to go through it again, step by step.

We begin by observing that neutrinos have masses that are very small, and possibly zero. Now, arguments based on the special theory of relativity imply that if a particle has zero mass, it must travel at the speed of light. Photons, for example, travel at this velocity. Naturally, this isn't very surprising, since photons *are* light. It is not known whether neutrinos have zero mass, or whether their masses are simply too small to be measured. However, this really has little effect on the argument I am outlining. If neutrinos have small, but finite, masses, then they would have traveled through the early universe very rapidly, at near-light velocities. The reason for this is very simple. At the time, the universe was very hot and large quantities of energy were available, and a given amount of energy will cause a light particle to move more rapidly than a heavy one. Thus particles such as protons and neutrons would have been traveling relatively slowly, while the much lighter neutrinos attained velocities that were very great.

The next step in the argument depends upon the observation that the expansion of the universe at this time must have depended upon the number of rapidly moving particles that it contained. The particles would have exerted a kind of outward pressure that would have influenced the expansion rate.

If there had been, say, four or five different kinds of neutrinos, the expansion would have been more rapid than it would have been had there been only three. And if the expansion had been more rapid, this would have produced effects that would still be visible today. In particular, the abundance of helium would be greater.

At this point, the argument becomes a bit technical. In order not to bog the reader down in too much detail, I will simply state that calculations indicate that the ratio of neutrons to protons in the universe depended, at this time, upon the expansion rate. If the expansion had been more rapid, the ratio would have changed in such a way that more helium would have been formed. When the calculations are carried out, and when we "plug in" the numbers for the amount of helium obtained today, we obtain the result that there are probably only three different kinds of neutrinos or, at most, four. The observed helium abundance is such that the existence of four neutrinos is just barely possible.

To the scientifically uninitiated, this argument may sound like some of the arguments in the *Analects* of Confucius. What I have in mind are so-called chain arguments that "prove" in numerous different steps that, if there is not order in the family, then the state is going to fall apart. Westerners tend to be skeptical about this kind of reasoning because they realize that, if any link in the chain turns out to be invalid, then the whole argument falls apart.* Can one really have faith in scientific arguments that seem to have the same character?

Most scientists would answer this argument with an emphatic yes. Scientific reasoning does often depend upon chains of ideas such as I have outlined above, but it differs from the chain arguments of Confucius in that each step can be tested experimentally. Scientists generally commit themselves to reasoning of this sort only after they have tested each individual link. Then, once a conclusion is reached, it is not forever enshrined as scientific dogma, but is tested experimentally, and if discovered to be incorrect, scientists then go back and try to discover if one or more of the links was not perhaps weaker than they thought.

In this case, it has tentatively been concluded that there are probably only three families of particles, and at most four. The next step will be to test this conclusion by performing experiments on the SSC to see if any direct or indirect evidence for additional particles can be found. If no evidence of this sort is discovered, then it will be concluded, at least for now, that the theory has been confirmed.

FERMIONS AND BOSONS

Even people who have no scientific knowledge are aware that force and matter are entirely different. The most obvious fact about matter, for example, is that it occupies space, and it is just as obvious that force doesn't. Yet if matter and force are both composed of certain fundamental particles, why shouldn't they be more alike? For example,

---

* When I use this example, I mean no disrespect to Chinese philosophy. The arguments of Western philosophers are often at least as suspect, and Western scientists have sometimes reached valid conclusions by means of arguments that contained outright fallacies.

why should light, which is made of photons, be so different from some such material object as a table, which is made of electrons and of up and down quarks?

The answer is that particles of force and particles of matter behave differently. There is a way in which they are fundamentally unlike each other, but this difference has nothing to do with mass, or with charge. On the contrary, it has to do with the particles' spin.

All the known force particles are bosons. Bosons, which are named after the Indian physicist Satyendranath Bose, have spins which are integral multiples of a certain fundamental unit. A boson, for example, can have 0 spin (0 is an integer, after all), or one unit of spin, or two.* Particles of matter, called fermions after the Italian physicist Enrico Fermi, have spins that are half-integral. A fermion, in other words, can have a spin of $1/2$, or $3/2$, or $5/2$,** or even something greater.

Bosons and fermions are unlike one another, because fermions obey Pauli's exclusion principle, while bosons do not. This principle, which is named after the Austrian physicist Wolfgang Pauli, says that no two similar particles with half-integral spins can occupy the same space at the same time. An electron cannot be squeezed into the space occupied by another electron, for example. With bosons, on the other hand, this can easily be done. In fact, to engage in a little anthropomorphizing, it is possible to say that bosons positively like being piled up upon one another. When this happens, the forces that they create simply become stronger.

These properties of fermions and bosons correspond exactly to the behavior that matter and forces exhibit in the

* The mathematically inclined will want to know that a unit of spin is defined to be $h/2\pi$, where h is a number called Planck's constant equal to 6.625×10^{-27} erg-seconds.

** Naturally, the latter two numbers could be written as $1\frac{1}{2}$ and $2\frac{1}{2}$, respectively. However, physicists prefer to write them as improper fractions.

everyday world. A table cannot be forced into the space occupied by another table, creating a single object that is twice as heavy. In the case of forces, on the other hand, this is exactly what we see. For example, two individuals pulling on a rope will exert twice the force exerted by one alone. Two beams of light can be superimposed on one another, making a single beam that is twice as intense. Similarly, the earth, which contains more matter than the moon, exerts a gravitational force that is correspondingly greater.

The Pauli exclusion principle does not explain why particles with half-integral spin should behave in this manner while those with integral spin do not. It is simply observed that they do. No exceptions are known.

Susy

In recent years, some physicists have begun to wonder whether the distinction between bosons and fermions doesn't sometimes break down. In particular, they have been exploring the implications of a certain theoretical idea known as supersymmetry. Since physicists seem to be enamored of cute nicknames these days, supersymmetry is sometimes called Susy (or susy) for short.

The basic idea of supersymmetry is exceedingly simple. It is postulated that there are not really two different kinds of particles, but only one. In supersymmetric theories it is assumed that every particle can be paired with another particle that is identical to it in every way, except that its spin and mass are different. Every spin-$\frac{1}{2}$ fermion, for example, is paired with a spin-0 boson. But this boson would not be any of the familiar ones that we have previously encountered; on the contrary, it would be a new particle that has not yet been seen in experiments.

In supersymmetric theories, the spin-$\frac{1}{2}$ electron has a partner called the selectron. Similarly, each of the spin-$\frac{1}{2}$ quarks would be paired with a spinless particle called a squark. Similarly, the spin-1 photon would be paired with a spin-$\frac{1}{2}$ photino (which would transmit force even though it was a fermion). Even the elusive Higgs particle would have a partner, the spin-$\frac{1}{2}$ Higgsino (if it exists, the Higgs particle would have to be a boson of spin 0).

It really isn't clear what the masses of the supersymmetric particles, which are collectively known as sparticles, should be, but it is clear that they would have to be very heavy. Otherwise, some of them would already have been seen in experiments performed on existing accelerators. For example, existing experimental data imply that, if the selectron exists, then it must be at least forty thousand times as heavy as the electron.

At this point, we might question whether it is really worthwhile to explore theoretical ideas, such as supersymmetry, for which no experimental justification exists. I think that the answer to this question has to be a conditional yes. Although it is true that there is no experimental evidence for supersymmetry, there are a number of respects in which it is nevertheless a very appealing idea. In the first place, it is known that the elementary particles exhibit various different kinds of symmetries. For every electronlike particle (electron, muon, and tau), there is a corresponding neutrino. Quarks come in pairs, and there seems to be a pair of quarks for every pair of leptons. If supersymmetry turned out to be a correct description of nature, everything could be tied together in a very appealing package, and the somewhat arbitrary distinction between particles with integral spin and those with half-integral spin could be broken down.

In other words, supersymmetric theories have a certain aesthetic appeal. This is not an irrelevant consideration. It has often turned out to be the case in the past that the most aesthetically pleasing theories have been the ones most likely to be true. Nature seems to be organized in simple, logical patterns, and some of the greatest discoveries in the history of science have taken place when scientists perceived these patterns and saw that they could explain physical phenomena in "beautiful" ways. Aesthetic considerations played a role in Galileo's preference for the Copernican theory of the solar system over the Ptolemaic, for example, and in Einstein's discovery of the ideas upon which his theories of relativity were based.

That a theory is beautiful does not necessarily imply that it is true. Numerous lovely theories have been disproved by ugly experimental facts. However, if we have to make a choice between a theory that possesses aesthetic appeal and one that does not, we rarely go wrong by picking the former.

Supersymmetry is not only a beautiful idea, but one that might point the way to that Holy Grail of particle physics, the unification of the forces. As we will see in chapter 8, the only theories developed so far that might lead physicists to this goal are those which incorporate supersymmetry. If any of these theories turn out to be true, physicists might finally find the answers to some of the questions with which the standard model has not been able to cope. A theory that incorporated supersymmetry, for example, might tell us why the forces observed in nature have strengths that are so different, and why the observed particles have the masses they do.

As we shall see later on, some of these theories are considered quite promising. Naturally they would be even more promising if there were some evidence that nature really is supersymmetric. This is where the superconducting supercollider comes in, for the experiments to be performed on the SSC could very well provide such evidence. If any gluinos (the supersymmetric partner of the gluon) or squarks have masses of less than about 1.5 TeV, the SSC will presumably be able to detect them, and evidence might very well turn up for the existence of other sparticles as well.

On the other hand, if no sparticles are observed, physicists will again be confronted with the problem alluded to in the last chapter, the fact that in many areas of physics, theory is beginning to outrun experiment. It is worth emphasizing once again that, if an idea has great imaginative appeal, that is no guarantee that it will turn out to be true. The human mind, after all, is capable of inventing an infinite number of different possible theoretical worlds. But, however plausible and appealing an idea might be, it is still necessary to perform experiments to find out if it corresponds to reality.

6

The Invisible Universe

ONE OF the most obvious and surprising things about the universe is that it is full of galaxies. Stars are not randomly distributed through space, but are found in huge galaxies of various shapes and sizes, most of which are large indeed. An average-size spiral galaxy such as our own Milky Way may contain about a hundred billion stars, while a giant elliptical galaxy can contain ten times as many. Even dwarf galaxies (such as the Large and Small Magellanic Clouds, which orbit the Milky Way) are collections of billions of stars.

Furthermore, galaxies themselves cluster together in groups. A typical cluster of galaxies might have anywhere between ten and a hundred members. The Milky Way, for example, is a member of a cluster called the Local Group, which also contains the great galaxy in Andromeda, and about twenty smaller systems. There are groups of galaxies that are much larger, and some superclusters contain as many as a thousand members.

It is not difficult to visualize how the galaxies were created. Billions of years ago, there must have been regions, hundreds of thousands of light-years across, in which the density of the primordial hydrogen and helium gas was greater than it was elsewhere. Gradually, these clouds contracted under the influence of gravity. After hundreds of millions of years, they broke up into fragments. The fragments contracted even more, forming regions of still higher density. As gravity compressed these clumps still further, they became hot, and nuclear reactions started up in their cores. One by one, the stars began to blink into existence.

There is only one problem with this scenario: according to the big-bang theory, it should never have happened. The expansion of the universe should have caused matter to become so dispersed that gravity should never have had a chance to collect it together, and the universe should be filled with tenuous hydrogen and helium gas, not with galaxies and stars.

Galaxies could only have formed if the condensations of matter from which they were created existed very early in the history of the universe. In other words, if regions of greater-than-average density were created soon enough, gravity would have won out over the expansion of the universe. However, this idea is contradicted by observations. If the early universe had contained regions of high density, then higher-than-average quantities of radiation would have emanated from these regions. If this had been the case, the effects would be visible today. The cosmic background radiation would not be as uniform as it is; there would be radio "hot spots" in the sky. Since these are not seen, we can only conclude that matter in the universe was still rather uniformly distributed when this radiation was emitted about half a million years after the big bang.

On the other hand, we know that matter must have condensed into galaxies relatively quickly. Galaxies, after all, are very old. All the evidence seems to indicate that they already existed within a few billion years after the big bang. The Milky Way contains stars thought to be 14 billion years old—almost as old as the universe itself.

Theory says that matter should not have condensed into galaxies. The observational evidence indicates that it condensed quickly, within the space of a few billion years at most.* Obviously, there is a contradiction here, one that must somehow be resolved.

The inflationary universe theories have solved some of the problems that have plagued cosmology. At first glance, however, it would seem that the inflationary paradigm would have little bearing on the problem of galaxy formation. After all, the inflationary expansion is supposed to have gone on for only a tiny fraction of a second, while the era of galaxy formation belongs to a much later epoch. As it turns out, the inflationary paradigm is not as irrelevant as one might think. As we shall see, the presumed existence of a period of inflationary expansion has implications for the quantities of matter that should exist in the universe. In particular, the inflationary theories say that there should be much more matter than there seems to be, that the bright, glowing matter that has collected into stars and galaxies is only a small fraction of the whole.

DARK MATTER

Astronomers have known for more than fifty years that there is something out there that they cannot see. The universe contains a mysterious kind of matter that telescopes cannot detect, but which nevertheless makes its presence known by

---

* A few billion years is obviously a long time compared, for example, to human life spans, but it is a short time on the cosmic scale.

exerting a gravitational pull on objects that astronomers can observe.

This effect was first noted by the Dutch astronomer Jan Oort around 1932. Oort had been studying stars that had moved away from the disk of our Milky Way galaxy. As these stars begin to climb above the disk, gravity acts to pull them back. As a result, they move more and more slowly, and eventually fall back in the direction from which they have come. By studying the positions and velocities of such stars, it is possible to calculate how much mass the galactic disk must contain.

When Oort added up the masses of the stars that could be observed in the disk, he found that there was apparently less than there should have been. The amount of matter that was visibly present was only about 50 percent of the quantity required to produce the observed motions.

Assuming that the discrepancy was caused by the presence of small stars too faint to be seen and counted, Oort added a correction to his equations to account for them. However, the discrepancy could not be banished for long. Later studies showed that these faint stars did not exist in sufficient numbers to cause the observed effects. Nor was the interstellar gas present in the galaxy able to provide the required mass. Something was exerting a pull on these stars, and that something could not be seen.

In 1933, a similar effect was pointed out by the California Institute of Technology astronomer Fritz Zwicky. Studying a large cluster of galaxies in the constellation Coma Berenices, Zwicky discovered that, although the galaxies in the cluster were obviously held together by their mutual gravitational attraction, the mass present in the stars that could be seen in the galaxies provided only a fraction of what was needed. As Zwicky put it, there was a "missing mass" problem.

Astronomers no longer speak of missing mass. Nowadays, they prefer the term dark matter instead, and indeed the latter term is more accurate, because there really isn't anything "missing." The problem isn't one of mass that should be there, but which isn't. The observed gravitational effects are caused by matter that obviously is present, but which astronomers cannot see. This material, incidentally, is called "dark" matter, not because it is dark in color, but because it gives off no light. It could just as well be called "invisible matter" instead.

Although more than half a century has passed since the existence of dark matter became known, astronomers are still not sure what it is. Nor are they certain precisely how much of it exists. They only know that there is quite a lot of it. According to current estimates, dark matter accounts for something between 90 and 99 percent of the mass of the universe.

During the 1980s, new techniques were developed that made it possible to detect the presence of large amounts of dark matter in and around galaxies. It was quickly established that galaxies, including our own, were surrounded by massive halos of some mysterious, invisible material. Again, the dark matter made its presence known by its gravitational effects. In order to explain precisely what these effects were, it might be helpful to digress briefly, taking a look at the similar, but simpler, problem of determining the mass of the sun.

Obviously, the sun cannot be placed on a balance and weighed. Offhand, one might think that it would be difficult or impossible to weigh it at all, but there exists an indirect, but very accurate, method of determining its mass. We need only observe the motions of the planets that orbit around the sun.

If the sun were more massive than it is, then the earth, Mars, Venus, and all the other planets would be whipped around by the sun's gravity at a more rapid pace. If the sun weighed less, their motion would be more leisurely. Calculating the sun's mass is a straightforward problem in Newtonian physics.*

---

* Although Newton's law of gravitation has been superseded by Einstein's general theory of relativity, it gives accurate enough results under many circumstances. Such is the case here.

Entire galaxies can be weighed in a similar manner. The stars in a galaxy orbit the galactic center. To be sure, the time of revolution is very great: for example, the earth revolves around the sun in a year, but it takes the sun 250 million years to complete one revolution around the center of the Milky Way. However, the principle is the same. The greater the gravitational force pulling an orbiting object, the greater its velocity will be.

To be sure, the mass of the sun is concentrated into one relatively small, spherical object, while the mass in a galaxy is distributed among billions of individual stars; but this really makes little difference. According to Newton's law of gravitation, it doesn't make any difference whether a gravitational mass is concentrated in one spot, or "smeared out" over a large volume. The motion of an orbiting star is determined by the quantity of mass *inside* its orbit.

I have italicized the word *inside* because this is an important point, since it means that we can use Newton's law to calculate the mass of a galaxy within any given radius. For example, if we look at the motions of the stars located two-thirds of the way from the galaxy's center to the galactic rim, then we can find the mass of that inner two-thirds of the galaxy.

It is also possible to use this method to measure mass that exists beyond the rim of the visible galactic disk. We need only measure the velocities of objects, such as clouds of gas, that follow orbits lying entirely outside of the galaxy. If we find that such clouds exist at varying distances from the galactic rim, it is not difficult to map the distribution of mass in a galaxy's dark halo. Nor is it difficult to carry out the necessary velocity measurements, since velocity is related to redshift. If we view a galaxy edge on, its rotation will cause the stars on one edge to move toward the earth while the stars on the other edge move away. It is something like viewing a rotating phonograph record edge on. In the case of the galaxy, the light from the stars that are approaching the earth will be blueshifted, while a redshift will be observed for those that are moving away. Naturally the overall redshift caused by the recession of the galaxy has to be taken into account, but this is easily done. Nor do any great problems arise if a galaxy is not seen exactly edge on, but at an angle. In this case, we need only note the angle at which the galaxy is being viewed, and use a little trigonometry to make the necessary corrections.

When these observations are carried out, it is found that considerable dark matter exists outside the visible galactic disks. In fact, the farther we go out from the center, the more mass we find. It appears that there is a great deal of invisible matter both inside and around galaxies. Dark matter is ubiquitous indeed.

WHAT DARK MATTER ISN'T

The problem of the nature of the dark matter has been widely discussed, and numerous exotic possibilities have been suggested. There would be little point in discussing such possibilities if the more obvious ones were not eliminated first. I will therefore begin by discussing what the dark matter apparently is *not*.

It isn't interstellar dust or some form of interstellar or intergalactic gas. Galaxies do contain dust clouds, and gas exists both within galaxies and in the spaces between them. However, dust and gas can be seen, and it is easily established that they could make up only a small fraction of the dark matter at best. Dust is easily visible because it obscures the light from more distant stars. Gas can be seen because it emits radiation: cold gas gives off radio waves, while hot gas emits ultraviolet radiation or X rays.

It so happens that the universe is full of X rays coming from all directions of the sky. However, it has been shown that most of these X rays come from other sources than clouds of gas. Quasars, those distant objects that are thought to be the luminous cores of newly formed galaxies, are a common source of X rays, for example. It is thus possible to place limits on the amounts of hot gas that can exist, and it turns out that there cannot be enough of it to account for the dark matter.

If the universe were full of chunks of rock, ice, or even "snowballs" of frozen hydrogen, these could provide an explanation for the dark matter. However, no one takes such possibilities very seriously. If such a theory were to be considered, it would be necessary to suggest a plausible explanation for the origin of such objects, and no one has ever come up with a reasonable idea to explain where they might have come from.

On the other hand, one could attempt to revive the suggestion that was originally made by Oort, and later discarded. The dark matter could be made up of tiny stars, ones that are too dim to be seen. For that matter, it could be made of objects, like the planet Jupiter, which might have become stars if they had only been slightly larger. Jupiter, by the way, can be described as a "failed star." Its composition, about 25 percent helium and 75 percent hydrogen, is practically identical to that of the sun. If Jupiter had been only slightly larger, gravity would have compressed the material in its core to such an extent that nuclear reactions like those that go on in the sun could begin. In that case, ours would be a binary star system.

Finally, it is possible that the dark matter could be composed of black holes. A black hole is the massive remnant of a dead star with a gravitational field so strong that nothing, not even light, can escape from it. It is not true, by the way, that it would be impossible to see a black hole. The intense gravitational fields that surround black holes draw matter into them, and this matter glows as it is accelerated. A black hole can be quite bright.

Although some objects have been observed that astronomers believe to be black holes, no one really knows how many exist in the universe. The details of star formation are still not completely understood, and until they are, it will be impossible to estimate how many stars might exist that are large enough to form black holes when they die. However, it seems improbable that there could be a sufficient number of black holes in a typical galaxy to account for the detectable dark matter. Furthermore, we would not expect to find a lot of black holes in the halos that surround galaxies, since there are few stars in such regions.

BARYONIC AND NONBARYONIC MATTER

There is no evidence, then, to indicate that the dark matter consists of dim stars, "Jupiters," or black holes. But if the dark matter is ordinary matter—matter made of protons, neutrons, and electrons—there are apparently no other possibilities. To be sure, matter that has collapsed into black holes can hardly be called "ordinary." However, black holes are formed from massive stars, which have chemical compositions similar to those of the other objects that we have been considering.

Up to now, I have been speaking of "ordinary" matter. However, the term *baryonic matter,* the one that is used by physicists, is really more precise. Baryonic matter is just that, matter made of baryons. In practice, this means neutrons and protons, since all of the other baryons are generally seen only in the laboratory. It is true that ordinary matter contains electrons, which are not baryons, also. However, the mass contributed by electrons is very small. The electron weighs only $1/1836$ as much as a proton. Electrons make up $1/1837$ of the mass of hydrogen atoms, therefore, and only about $1/3675$ of helium.

In recent years, scientists have discovered powerful theoretical arguments that seem to imply that the greater part of the mass present in the universe must be nonbaryonic. If these arguments are valid, then the dark matter is not made of protons and neutrons (and electrons) at all. It may be composed of exotic particles not yet seen in the laboratory, or even of objects that cannot properly be called "particles" at all.

In chapter 3, we saw that observations of the quantities of helium and deuterium that are present in the universe provide an important confirmation of the big bang theory. I pointed out that the observed quantities of helium are far greater than the amounts that could have been made in stars. The deuterium that is seen could not have been manufactured in stars at all; it could only have been created in the big bang.

We also saw that measurements of helium could be used to make inferences about the number of different neutrino varieties that can exist (see chapter 5). It should come as no surprise, therefore, if we discover that measurements of deuterium abundance have important implications also.

This is indeed the case, once we realize that only some of the deuterium that was created in the big bang still exists today. Many of the deuterium nuclei that were created in the big bang fireball must soon have collided with one another to form helium. Thus the abundance of deuterium today is related to the density of matter in the early universe. The denser the universe was then, the greater the number of collisions that would have taken place. Thus observations of deuterium abundance today should tell scientists how dense the universe was back then.

However, if we know how dense the universe was when it was a few minutes old, we can calculate how dense it should be now. Calculations based on measurements of deuterium in the present-day universe show that the density of baryonic matter cannot be more than 20 to 30 percent of the critical density required to close the universe. The calculation only tells us how much baryonic matter should be present, naturally. After all, a deuterium nucleus consists of a proton and a neutron, and both of these particles are baryons.

Similar calculations based on the observed abundances of helium 3 and lithium 7 can be carried out. Helium 3 is a form of helium in which there are three particles in the nucleus, two protons and a neutron, rather than the usual four (two protons and two neutrons). Lithium 7 is a form of the metal lithium in which there are seven particles in the nucleus, three protons and four neutrons. Both these substances are present in even smaller quantities than deuterium. Lithium 7, for example, contributes only about one ten-billionth of the total mass of the universe. However, such abundances can be measured. Calculations show that the density of baryonic matter is between 3 and 12 percent of the critical density. This result is in agreement with that obtained from deuterium measurements since the figure of 20 to 30 percent cited above is only an upper limit.

When calculations of this sort were first performed during the 1960s, scientists drew what then seemed to be the natural conclusion. If the matter density was well below the critical value, then the universe had to be open; it was fated to expand forever. It never occurred to anyone that there might exist nonbaryonic forms of matter that contributed significantly to the total mass of the universe. At the time, nonbaryonic matter was not even considered to be a theoretical possibility.

Since that time, matters have changed considerably. New theories in the field of particle physics imply that there might exist numerous different kinds of particles that have not yet been seen in the laboratory. If these particles really exist, they might contribute significantly to the total mass density of the universe. Furthermore, the inflationary universe theories seem to imply that the overall mass density is much greater than the maximum of 20 to 30 percent implied by the deuterium argument. If this is the case, the greater part of the matter in the universe has to be nonbaryonic.

In fact, if an inflationary expansion did take place, the mass density of the universe must be equal to the critical value, or be so close to it that we may never be able to measure the difference. As we saw in chapter 4, the inflationary theories require that the universe be on the borderline. It may be just barely open, or just barely closed. In either case, the mass density would have to have a borderline value also.

In an inflationary universe, then, at least 70 percent of the matter has to be nonbaryonic, though nonbaryonic matter could easily account for 90 percent of the mass in the universe, or even more. The luminous mass in the universe—that which exists in stars and in other glowing objects—is only about 1 percent of the critical density.

NEUTRINOS

For about fifty years after the existence of the neutrino was postulated in 1930, physicists assumed that this particle had zero mass, and that it traveled at the speed of light. There was really no particular reason why the mass had to be zero, however. This assumption was simply retained in the absence of any evidence to the contrary.

But then, around the beginning of the 1980s, two experiments were performed that suggested that the neutrino might have a small mass after all. It was immediately apparent that, if this was indeed the case, neutrinos could be the dominant form of matter in the universe. Since there were about a billion neutrinos for every baryon, they would weigh more than everything else combined if their mass was just a fraction of the electron mass.

In 1980, three University of California at Irvine physicists, Frederick Reines, Henry W. Sobel, and Elaine Pasierb, reported that they had observed neutrino oscillations. In other words, they had found that one variety of neutrino could change into another. An electron neutrino might be transformed into a muon neutrino, for example, and then change back into an electron neutrino again at a later time. Or—and it was this particular phenomenon that the three experimenters thought they had detected—oscillations might take place between electron and tau neutrinos.

According to currently accepted theory, neutrino oscillations can take place only if neutrinos have mass. The Reines-
Sobel-Pasierb experiment didn't say how large this mass
should be, but it did seem to imply that it could not be zero.

The experiment involved many uncertainties, not the least
of which was the fact that the tau neutrinos were never detected. When some of the electron neutrinos disappeared for
a while, it was simply assumed that they had been transformed into the tau variety. Thus the experimenters could not
claim that their results were conclusive, and agreed that the
results would have to be confirmed by other experiments.

Though no experimental confirmations were immediately
forthcoming, the result still attracted a lot of attention. There
were a number of reasons for this. One was the fact that the
GUTs seemed to imply that neutrinos should have mass.
Another was the prestige of one of the experimenters. In
1956, in collaboration with another physicist, Clyde L.
Cowan, Jr., Reines had performed the experiment in which
the existence of the neutrino had finally been demonstrated,
twenty-six years after its existence had first been proposed.

Interest in the possibility of neutrino mass subsequently
increased when a group of scientists at the Institute for Theoretical and Experimental Physics in Moscow reported that
they had measured the mass of the electron neutrino directly,
and had found that it was somewhere between 14 and 48 eV.

The Soviet result was not widely accepted. The experiment
had been a difficult one, and the quantity that had supposedly been measured was very small. Nevertheless, theoretical physicists all over the world went to work figuring out
what the implications of neutrino mass would be. They
quickly found that, if the mass of the electron neutrino was
even 2 eV, then the combined weight of all the neutrinos that
existed would be greater than that of all of the baryonic
matter in the universe. Furthermore, if the neutrino weighed
14 eV, then these particles would contribute 90 percent of the
universe's mass.

Furthermore, if neutrinos had mass, they would not be

able to travel at the speed of light. And if this was the case, then they could be slowed to velocities that would allow them to be captured by the gravity of galactic clusters. The dark halos that surrounded galaxies could easily be made of neutrinos. For that matter, there seemed to be no reason why concentrations of neutrinos could not have been responsible for the formation of galaxies in the first place. If gravity had caused neutrinos to cluster together shortly after the big bang, ordinary matter might later have accumulated in clumps of neutrinos. Gravitational attraction would see to that. Since neutrinos did not emit light, the uniformity of the cosmic background radiation would not be a problem.

For a short time, it appeared that scientists had gone a long way toward solving the problem of galaxy formation, but difficulties soon began to crop up. The experimental results which had indicated that neutrinos possessed mass were not confirmed, and scientists became skeptical of these results. It became apparent that no one could really say whether neutrinos had mass or not.

There were theoretical difficulties too. Since neutrinos are very light particles, it was obvious that they would have emerged from the big bang at velocities close to that of light, accelerated by the energy that was available. But such free streaming neutrinos could not have created mass concentrations in the universe; they would have broken up mass concentrations instead. In other words, if any galaxy-size clumps of matter had somehow been formed, the neutrinos would have scattered them apart.

Only after the neutrinos had slowed to velocities about one-tenth the speed of light could they have begun to clump together. But calculations indicated that, if such a process did take place, neutrinos would have formed mass concentrations the size of superclusters of galaxies. The superclusters would then have had to break up into individual galaxies.

This scheme is called the top-down scenario for galaxy formation. Large mass concentrations are created first, and smaller ones follow later. Though it might at first seem reasonable that galaxies could have been created in this manner, there is one serious problem: The time required for galaxy formation is too long. Computer simulations indicate that up to four billion years are required to complete this process. Yet there is evidence that galaxies already existed only two billion years after the big bang. It seems we must conclude that the hypothesis that neutrinos have mass does not solve the problem of the existence of dark matter.

HOT AND COLD DARK MATTER

Particles, such as neutrinos, which emerged from the big bang at high velocities, and which might have clumped together in this manner are called hot dark matter. Here, the term *hot* has nothing to do with the overall temperature of the universe at the time. It is simply a reference to the fact that the particles would have been moving rapidly.

Cold dark matter, on the other hand, would be made of particles that emerged from the big bang at relatively low velocities. Again, the word has nothing to do with the overall temperature of the universe, but an analogy can be made with the way in which molecules in a hot object move about rapidly, while those in a cold object move at a much slower pace. Cold dark matter particles would be much heavier than particles of hot dark matter, for the simple reason that heavy particles have more inertia and are thus harder to accelerate.

Physicists sometimes speak of the particles that might make up cold dark matter as WIMPs, or weakly interacting massive particles. WIMPs have not yet been observed to exist. However, as we saw in chapter 5, there are theoretical reasons for suspecting that a number of different kinds of as-yet-unseen particles might exist, and scientists hope that they will be able to create some of them in experiments on the SSC.

In the meantime, there is nothing wrong with engaging in

a little speculation and trying to see where the hypothesis of the existence of cold dark matter would lead. Consequently, cosmologists have attempted to see if this type of matter could produce galaxies and galaxy clusters of the observed sizes.

In one respect, the cold dark matter hypothesis turns out to be very successful. Since WIMPs would have emerged from the big bang at low velocities, there would have been little or no free streaming, and mass concentrations would not have been dissipated. In fact, matter could have begun to have clumped together relatively quickly. Small mass concentrations would have formed first. Larger aggregates, such as clusters and superclusters of galaxies, would have formed later. In this bottom-up scenario, galaxies would have been created relatively quickly.

Unfortunately, like the hot dark matter hypothesis, the cold dark matter theory also runs into problems. The assumptions upon which the theory is based seem to lead to the prediction that galaxies of about the right size would have formed at about the right time. However, the theory doesn't seem to be capable of explaining the large-scale structures observed in the universe. Astronomers have discovered that galaxies and clusters of galaxies seem to be grouped together in long chains and filaments, and that there are huge voids, measuring as much as 250 million light-years across, in which few or no galaxies are found. But the cold dark matter hypothesis seems to imply that galaxies should be distributed throughout the universe in a more or less random way.

Neither the hot dark matter nor the cold dark matter hypothesis seems to work, at least not in its pure form, but this doesn't necessarily imply that these ideas have to be completely discarded. It is possible that some modification of one scenario or the other might produce acceptable results. Indeed, one such modification will be discussed shortly. First, it might be well to consider one of the alternatives to the hot- and cold-dark-matter-theories.

SHADOW MATTER

In chapter 8, I will discuss a group of theories, called super-string theories, which are currently creating a great deal of excitement in the theoretical physics community. Since they will be described in detail later on, there would be little point in saying very much about them here. I should mention, however, that some of these theories predict the existence of a strange substance called shadow matter, which would interact with ordinary matter only through the gravitational force. This implies that it could be neither seen nor felt.

Shadow matter could not be seen because light is a form of electromagnetic radiation, and a substance that did not feel the electromagnetic force would neither emit nor reflect light. Shadow matter could not be felt because electromagnetism is also responsible for the forces that hold atoms and molecules together. If someone tried to grasp a chunk of shadow matter, her hands would pass right through it.

It has been said that one would walk through a shadow matter mountain or stand at the bottom of a shadow matter ocean and never know it. However, even if shadow matter is real, such things probably do not exist. To be sure, shadow matter particles could interact with one another according to physical laws similar to those of our world. It is just barely possible that there could be shadow matter stars and shadow matter planets, and perhaps even shadow matter organisms, but it is probably more likely that the laws of nature would be different in the shadow matter world, so different that none of these things would be created.

If shadow matter does exist, it might consist of nothing more than clumps of particles similar to the matter concentrations that could presumably be produced by hot or cold dark matter—but at this point, even this is only a bizarre possibility. If scientists discuss the possibility of shadow matter, it is not because there are any good reasons for thinking that it is real; as yet, there are not. Scientists discuss it because it is necessary to consider every possible idea if we are ever to be sure precisely what the dark matter in our universe is.

COSMIC STRINGS

Another possibility is that the dark matter might consist, at least partially, of cosmic strings. Cosmic strings, which have nothing to do with superstring theory in spite of the similarity in the names, are cracks in the structure of spacetime that might have been created, according to some of the GUT and supersymmetry theories, when the universe was about 10^{-35} seconds old.

A cosmic string would be a discontinuity created when the quantum fields in the early universe underwent sudden changes. It would bear some resemblance to a flaw in a diamond, or to a crack that might appear on the surface of a frozen lake. If cosmic strings exist today, they would have the form of long, filamentlike concentrations of energy. It should be emphasized, by the way, that scientists have no evidence that cosmic strings exist. All that they can say is that their existence is somewhat less speculative than that of shadow matter.

If cosmic strings exist, they must be very massive. A piece of string the size of an atom would weigh a billion tons, and a section long enough to stretch across the width of a football field would weigh as much as the earth. Thus cosmic strings could easily have played an important role in galaxy formation, since their great mass would certainly have produced concentrations of matter.

We cannot conclude, however, that cosmic strings might constitute the dark matter that exists today. Most strings would have evaporated by now. The theory predicts that cosmic strings would vibrate quite rapidly, and that their energy would be radiated away. Thus the strings would evaporate; after all they are nothing but energy. The smallest strings would disappear most quickly, while larger ones could survive somewhat longer.

Even though few cosmic strings would still exist today, they could nevertheless explain why galaxies and clusters of galaxies have the forms they do. In fact, the hypothesis of the existence of cosmic strings seems capable of "saving" the hot dark matter theory. If there were strings in the early universe, neutrinos could have clustered around them. The extra gravitational attraction created by the strings would have lead to galaxy formation at an earlier time than would have been possible with neutrinos alone.

In a way, the theory seems a little too good to be true, since it holds that galaxies were supposedly produced by cosmic strings, which then conveniently radiated away into nothing once they had done their job. Belief in the theory would be considerably strengthened if some evidence for the existence of cosmic strings could be found, which would make it a little easier to believe that the strings had once been numerous.

Finding such evidence may not be as hopeless a task as you might think. Since the largest strings would live the longest, it is entirely possible that some strings were massive enough to survive to the present time. And if such strings exist, there are a number of ways in which their presence might be detected. For example, the gravitational fields they created would bend any rays of light that happened to pass by, since, according to Einstein's general theory of relativity, light is affected by gravity, an effect that has been observed on numerous different occasions.

The presence of a cosmic string could presumably be detected through a gravitational lens effect. If a string were positioned between the earth and a distant quasar or galaxy, light from that object could bend around both sides of the string as it traveled to the earth. Astronomers would then see two or more images of the galaxy or quasar instead of one.

Cosmic strings would also emit gravitational radiation. This would be the analogue of the electromagnetic radiation associated with the electromagnetic force. Since gravity is a much weaker force than electromagnetism, gravitational waves would be weaker and harder to detect than such forms of electromagnetic radiation as light, radio waves and X rays. Scientists have not yet been able to devise any experimental means for verifying the existence of such gravitational waves.

However, the level of experimental technology is constantly rising, and new ideas are constantly being proposed. There is every reason to think that gravitational radiation will be observed before too many years pass. Furthermore, indirect means of detecting the gravitational waves that might be produced by cosmic strings have already been suggested.

In particular, calculations indicate that the gravitational radiation produced by cosmic strings would have observable effects on the behavior of pulsars. Pulsars are collapsed, rapidly rotating stars that emit radio waves or other radiation in equally spaced pulses. Because pulsars rotate, the beams of radiation that they emit sweep past the earth like the beam of a searchlight sweeping past a stationary object. Gravitational radiation from cosmic strings could cause the timing of these pulses to become somewhat irregular. At this writing, experimental technology is not yet accurate enough to detect such irregularities. However, there is every reason to think that there will be sufficient improvement in the future that the experiment can be carried out.

DOES DARK MATTER REALLY EXIST?

Yes, it does. This has been firmly established. Nonluminous mass has been detected in numerous places in the universe. In fact, it has been shown that at least 90 percent of the mass of galaxies resides in the dark halos.

On the other hand, the question, "Does nonbaryonic dark

matter really exist?" cannot be answered so simply, since it is reasonable to assume that at least some of the dark matter that has been detected is baryonic. It could exist as dim stars, as "Jupiters," as black holes, or even as failed galaxies (a failed galaxy would be a large clump of baryons that condensed gravitationally but which didn't produce stars—no one knows whether or not failed galaxies really exist).

The density of luminous matter in the universe is about 1 percent of the critical density. There could be ten times as much dark baryonic matter. However, if the theoretical arguments based on observations of the abundances of deuterium, helium 3 and lithium 7 are correct, the total amount of baryonic dark matter cannot be much more than about 10 percent of the critical density.

There is only one reason for believing in the existence of nonbaryonic dark matter: Its existence is predicted by the inflationary universe theories. If the inflationary paradigm is correct, then the mass density of the universe must be very close to the critical density. The limits on the density of baryonic matter would then imply that the universe was mostly nonbaryonic.

At this point, it is worth pointing out, once again, that the inflationary paradigm has not been experimentally verified. Although it has become part of standard cosmological theory, there is little or no observational evidence to support it. Its wide acceptance can be attributed to its plausibility and its apparent ability to explain a wide variety of phenomena, not to its agreement with experimental results.

It should be noted, however, there exists no other theory that even comes close to explaining so many of the observed features of the universe. We can hardly go wrong, therefore, by making the provisional assumption that there was indeed a period of rapid inflation that began when the universe was about 10^{-35} seconds old. No one has come up with any better ideas.

As we have seen, if such an inflationary expansion did

take place, it is difficult to avoid the conclusion that at least 90 percent of the mass of the universe resides in a form of matter that is something other than the familiar baryonic variety. No one yet knows what this matter might be, but we can assert with a reasonable degree of confidence that it exists.

7

The Farthest Things in the Universe

IF COSMIC strings exist, they would have a tendency to twist around and intersect with one another as they moved around in the universe. Eventually, they would break up into closed loops that would behave as gravitational "seeds" around which galaxies could form. Then they would radiate away their energy and disappear. At least, this is one scenario for galaxy formation, though not the only one that has been suggested. According to another theory, cosmic strings might be capable of producing enormous explosions that would blow matter away. These explosions would give rise to expanding bubbles of gas, and galaxies would be formed when the bubbles collided.

The theory had its origin in 1985 when Princeton University physicist Edward Witten suggested that cosmic strings might behave as superconductors. According to Witten's theory, the properties and behavior of subatomic particles might change when they were trapped inside strings. In particular, some particles might have no mass under such conditions. If this were the case, little or no energy would be required to create them.

If a pair of charged particles such as an electron and a positron were created within a length of string, and if these particles had no mass, then a very small amount of energy could cause them to move at the speed of light. They would have to move at that velocity because special relativity requires that massless particles travel at that velocity.

If an electron and a positron traveled around a loop of string in opposite directions, a net electric current would be created, which would be exactly twice that created by the electron or the positron alone, since the current created by a positive charge moving in one direction is exactly equal to that created by a negative charge moving the other way. This is analogous to the rule of arithmetic that says that subtracting a negative number is equivalent to adding a positive one (for example, subtracting -5 is the same as adding $+5$), or to the rule of English grammar which states that two negatives make a positive.

Once such a current were set up, no additional energy would be required to maintain it. A loop of cosmic string could, according to Witten's theory, behave in a manner similar to the superconducting materials upon which scientists experiment in terrestrial laboratories.

If a superconducting current were set up within a cosmic string, electric and magnetic fields would be created in the space around the string. These fields could then travel away from the string as electromagnetic radiation.

According to a theory developed by Witten in collaboration with his Princeton colleague Jeremiah P. Ostriker and the latter's student, Christopher Thompson, this could very well have led to the formation of galaxies. The electromagnetic waves emanating from strings would have interacted with the hydrogen and helium gas that filled the universe to produce expanding bubbles of hot gas. Galaxies could have been formed when the bubbles intersected.

It should be emphasized that, if these ideas are correct, such events must have taken place on a truly enormous scale, since there are many galaxies with masses a trillion times greater than our sun, and astronomers have discovered voids in space that measure millions of light-years across. If the Ostriker-Witten-Thompson theory is correct, every one of these voids could have been created by a single superconducting loop of cosmic string.

Although the theory seems capable of explaining the existence of voids as well as galaxies, this hardly guarantees its correctness. The voids could easily have been created in some other manner, and they did not always have the enormous dimensions they possess today. The voids expand along with the rest of the universe; at one time they were considerably smaller than they are now. They might even have begun as random fluctuations in the density of matter. If clusters of galaxies could form in regions where the matter density was unusually high, there could have been voids in regions where the density was unusually low.

If the Ostriker-Witten-Thompson theory is correct, magnetic fields must have existed in the universe during the period that preceded galaxy formation. Superconducting currents could have been set up in the cosmic strings only if there were magnetic forces acting on the particles within them. The theory does not explain how these fields originated; we must simply assume that they were there.

If this theory is correct, then the superconducting cosmic strings might still be visible today, even though they presumably would have evaporated billions of years ago. When the superconducting currents reached a maximum, the strings would have emitted copious quantities of radiation, which would still be visible in the form of X rays today. There are numerous other objects in the universe that emit

X rays, so scientists would have to establish that any observed X-ray sources emitted the right kind of radiation in the right quantities.

There is another way that the theory could be tested, which has to do with the magnetic fields found within galaxies. These intragalactic fields are not related to the primordial magnetic fields required by the theory, but would exist whether there was magnetism in the early universe or not. Rotating galaxies generate magnetic fields through a kind of "galactic dynamo" effect. The strength of such a field is typically about a millionth of the intensity of the earth's.

If any superconducting strings existed today, they could interact with these galactic fields. This would not produce explosions; these fields are too weak. But radio waves would be emitted, and these presumably could be detected. Again, there are numerous sources of radio waves in the universe. However, if astronomers discovered a source of radio waves, and could not explain how they had been created, the presence of a superconducting string would at least have to be considered as a possibility.

Like the light that comes from stars and galaxies, radio waves from a particular kind of source generally have a character specific to that source. For example, the existence of cold gas can be detected because it emits radio waves at certain specific wavelengths. Consequently, if some new source of radio waves were discovered, it should be possible to tell whether or not they were produced by some familiar kind of astronomical object. If they were not, then other explanations, such as the existence of a superconducting string, would have to be considered.

It should be emphasized, however, that until there is some evidence for cosmic strings—either superconducting strings or those of the "ordinary" variety—theories that depend upon their existence have to be regarded as quite speculative. Though it is interesting to see what consequences the existence of cosmic strings might have, they are, at present, nothing more than a fashionable idea. There is no evidence whatsoever for their existence in reality.

I am not implying, however, that such speculation should not be pursued. In the past, discoveries of great importance have been made when scientists were "playing" with new theoretical ideas. Furthermore, even wild speculation can play a useful scientific role in that it broadens scientific horizons. On the other hand, we should not make the mistake of confusing scientific speculation with well-established fact. It is perfectly possible that evidence for the existence of cosmic strings will never be discovered and that, in a few years' time, strings will have been forgotten. Science has its fads, too. Many of them eventually go the way of the Hula-Hoop.

BALLS OF WALL

Cosmic strings—if there really are such objects—can be thought of as "cracks" in space and time. These cracks presumably appeared when the quantum fields that existed in the universe underwent sudden changes. Until recently, scientists had assumed that such changes, called phase transitions, took place when the universe was only a tiny fraction of a second old. The very concept of cosmic strings, in fact, was originally developed in the context of the inflationary universe paradigm, and inflation was supposed to have come to an end about 10^{-30} seconds after the big bang. A phase transition, after all, is a dramatic event, and it is natural to think of it as something that took place amid violent events during the period of inflationary expansion. However, there is really no good reason why a phase transition could not have taken place at a later time. Quantum fields, such as the Higgs field and those associated with various particles, must have existed before the inflationary expansion began, and they were presumably still present after it ended.

In 1988, University of Chicago astrophysicist David N.

Schramm and his collaborators, Christopher T. Hill of the Fermi National Accelerator Laboratory and J. N. Fry of the University of Florida, suggested that a late phase transition did indeed take place about a million years after the big bang.

Schramm and his colleagues began by observing that the electron neutrino could easily have a mass of about 0.01 eV. Such a small quantity could not be measured. However, if the neutrino did have such a mass, this would clear up one of the outstanding problems in physics.

It so happens that the number of neutrinos coming to the earth from the sun is smaller than the number that should theoretically be observed. This discrepancy would be eliminated if the neutrino had some mass. As I have observed previously, if neutrinos had mass, neutrinos could oscillate from one variety to another. If they did, an experiment that detected only electron neutrinos would give lower-than-expected results. The electron neutrinos that had oscillated into muon neutrinos, for example, would simply not be seen.

If neutrinos have mass, then it is reasonable to assume that they acquire this mass through a process resembling the Higgs mechanism, but there is no particular reason why this mechanism must have given mass to neutrinos from the very beginning. This might not have happened until the universe was a million or so years old. According to Schramm, Hill, and Fry, it is conceivable that, at this time, a phase transition took place in which a Higgs-like field suddenly "froze," giving mass to these previously massless particles.

If such a phase transition took place, it is reasonable to think that it would have created "cracks" in spacetime similar to those that might have appeared during the period of inflationary expansion. However, there is no reason why the spacetime flaws created when such a "freezing" takes place have to be one dimensional, like cosmic strings. It is also possible to have pointlike defects (which would look like massive particles), or two-dimensional domain walls.

In fact, both these kinds of flaws could have been created during the period of inflationary expansion. The particles, called magnetic monopoles because they would behave as isolated north or south magnetic poles, would be very rare. The rapid inflationary expansion would sweep most of them out of the observable portion of the universe. As we saw in chapter 4, domain walls would experience the same fate. Presumably, they exist somewhere in the universe, but are so far away that we can't see them.

Cosmic strings would also have been swept out of the observable universe. This creates a problem for dark-matter theories that depend upon the existence of strings. There should not be enough of them to account for the formation of galaxies, unless they were created after the inflationary expansion had ended. This raises the question of why they should have been formed at this time, if the magnetic monopoles and domain walls were created earlier. Although the problem might not prove to be insurmountable, it would eventually have to be solved if a cosmic string theory were to be taken seriously.

Difficulties of this sort are avoided, however, if one postulates that a late phase transition took place. Any spacetime flaws created after inflation had ended would move along with the now leisurely expansion of the universe. They would not be swept away and disappear over cosmic horizons.

According to Schramm, Hill, and Fry, a late phase transition could very well have led to the creation of domain walls which later broke up and became the seeds for galaxy formation. However, such pieces of wall would bear little resemblance to the strings of the theories that we considered previously. While the strings would be microscopic structures many times thinner than the diameter of an atomic nucleus, the domain walls of late phase transition theory would be large structures indeed. Depending upon the size of the neutrino mass that was created by the Higgs-like field, they could be as much as millions of light-years thick. Naturally, they would not have the enormous mass density attributed to strings; most likely their density would be of the same order of magnitude as that of the hydrogen and helium gas in the surrounding universe.

There are two ways in which the creation of domain walls could have led to galaxy formation. First, a wall would exert a repelling, "antigravity" force on nearby matter. Matter between a pair of walls would therefore be compressed. Galaxies could thus be created in a manner similar to that postulated by the adherents of the exploding string theory. In either case, two bubbles of compressed matter would come together, and chains of galaxies would be created where the bubbles happened to intersect.

The other possible method of galaxy formation is a consequence of the fact that domain walls would not be hard, rigid objects. Deformations could take place in them, and pieces could break off and form balls of wall. Such domain-wall bubbles would exert an attractive gravitational force on the matter around them, and could seed galaxy formation just as loops of cosmic string could.

The late phase transition theory is very speculative. As yet, there is no evidence that neutrinos do indeed have mass, and there is certainly none that would indicate that a phase transition of the required character really did take place. However, the theory does have certain advantages over its rivals. If it is correct, then galaxies would form relatively quickly. The problems encountered in some other theories, where galaxy formation might be too slow a process to be consistent with observation, would be avoided.

The Schramm-Hill-Fry theory also avoids problems having to do with the homogeneity of the cosmic background radiation. Late phase transitions would take place after this radiation was emitted, and they would have little effect on it as it proceeded through space. Thus if the domain walls or balls of wall did produce any unevenness in the radiation, it would be too small to be observed today. The theory thus has an advantage over theories that depend upon the existence of density fluctuations at an earlier time. As we have seen, if these fluctuations were very large, their effects would still be visible today; a "lumpiness" in the universe before the background radiation was emitted would cause a lumpiness in the background itself.

Astronomers have observed no domain walls or balls of wall in the universe, but this doesn't really contradict the late phase transition theory. One can simply assume that, during the billions of years that have passed since the late phase transition took place, the walls and their remnants have dissolved away.

We might be a little skeptical of a theory that depends on the existence of objects that no longer exist. However, there may be ways that the theory can be tested after it is worked out in more detail. The theory might be used to make quantitative predictions about large-scale structure in the universe. For example, if it turned out that the theory predicted the existence of voids that were approximately the size of those that are actually observed, we would have to take it seriously. Similarly, if observational techniques were refined to the extent that it was possible to measure fluctuations in the cosmic background radiation that were much smaller than those that can be observed today, this might also provide a confirmation. The theory does predict small nonhomogeneities in the background, and says that they should have a certain specific magnitude.

However, at present, the late phase transition theory probably should not be viewed as anything more than an alternative to standard models of galaxy formation. If it avoids certain problems that they encounter, that is no guarantee that it is true, or even especially plausible.

Perhaps the best way to sum up the current situation is to cite some comments that Princeton University astrophysicist P. James E. Peebles made in the magazine *Science News* about the late phase transition theory. Peebles was quoted in the April 29, 1989 issue of *Science News* as saying,

Maybe something crazy is needed. None of the standard models for formation of galaxies and clusters of galaxies fits very well with all of the data. That could be because we're missing some elementary point in the way we approach the data, or it could be we're missing something big, like a late-time phase transition. I certainly wouldn't dismiss [that possibility], because we're getting a little desperate.

THE GREAT ATTRACTOR

Since the time that Galileo first used a telescope to observe the heavens, astronomers have always attempted to see as far out into space as they possibly could. Like all scientists, they were eager for new knowledge, and this seemed an obvious way to obtain it.

By pushing observational techniques to their limits, astronomers have been able to gain an understanding of the large-scale structure of the universe. As they have looked farther and farther into space, they have discovered new astronomical objects and phenomena, and have found themselves looking farther and farther back in time. Today, for example, astronomers can see more than ten billion light-years into space, and have thus been able to observe the universe when it was more than ten billion years younger.

One of the things that has made this possible is the development of new electronic technology that has enabled astronomers to make observations that were impossible in Hubble's day. Astronomers no longer confine themselves to observing the universe through optical telescopes that gather visible light. Nowadays they make use of every part of the electromagnetic spectrum, and observe the universe in the radio, infrared, ultraviolet, and X-ray bands as well.

When astronomers do use telescopes to expose photographic plates, they no longer need examine the plates visually. Laser scanning devices can read off the information contained in the plates in an instant. More often than not, astronomers don't use photographic techniques at all, but use modern electronic instruments instead. These advances seem especially impressive when we consider the fact that, in Hubble's day, the redshifts of distant galaxies had to be determined by crude "brute-force" methods. In those times, the exposure required to find the redshift of a single galaxy often lasted several days. The shutter of the astronomical camera would be closed during the day, and opened again the following night. Naturally, care had to be taken that the telescope was positioned exactly the same way on each successive night, or several nights of work might be wasted.

As astronomers attempted to look farther and farther into space, they tended to pay little attention to the data that had been collected on nearby galaxies. Some astronomers noted that the correlation between the distances of these galaxies and their redshifts was not as good as it should have been. However, preoccupied with other, "deeper," questions, the majority of their colleagues paid little attention to this fact. Those who did note it didn't seem to consider it very disturbing. The existence of these discrepancies, it was assumed, was nothing more than a reflection of the difficulties involved in measuring the distances to other galaxies accurately.

Few astronomers realized that another explanation was possible, that the galaxies were not simply being carried along with the general expansion of the universe, that they had peculiar motions of their own. Here, *peculiar* is used in the sense of "special" or "particular" rather than "odd" or "curious." A peculiar motion of a galaxy would be one that was caused by the gravitational attraction of concentrations of matter in the galaxy's vicinity. Such peculiar motions are observed, for example, in the Local Group. Some of the galaxies within this cluster are approaching, rather than receding from, the Milky Way. The reason is, of course, that the galaxies within the Local Group are gravitationally bound to one another.

Though astronomers realized almost at once that gravity was holding the Local Group together, it apparently never occurred to the majority of them that the universe might contain concentrations of matter that could affect the motions of galaxies on a larger scale. At any rate, for more than forty years after Hubble's announcement in 1929 of his discovery that the universe was expanding, the question was neglected.

When the problem of peculiar velocities finally was considered in the early 1970s, astronomers generally concluded that such velocities were probably small. There might be some random motions, it was argued. However, if they were very large, then many nearby galaxies would show blueshifts rather than redshifts. Since this was obviously not the case, it followed that there could be no significant irregularities in the expansion of the universe.

This consensus was upset before it had a chance to harden into scientific dogma. In 1975, astronomers Vera C. Rubin and W. Kent Ford, Jr., of the Carnegie Institution of Washington announced that they had determined that our galaxy had a velocity of about 500 kilometers per second with respect to the frame of reference of distant galaxies. The velocity was far larger than astronomers thought possible, and so the result was not widely accepted. It was pointed out that Rubin's and Ford's measurements depended upon finding groups of reference galaxies on opposite sides of the Milky Way that were about the same distance from the earth. To be sure, the critics said, the Milky Way *seemed* to be moving toward one group, and away from the other, but this could be an illusion that was caused by errors in distance estimates. There was simply no way of knowing whether or not the reference frame that Rubin and Ford had selected was a reasonable one.

Then, in 1977, it was discovered that the Milky Way *was* moving. It was established that it was in motion with respect to a frame of reference that everyone knew was reasonable, the cosmic microwave radiation background. Instruments sent aloft on balloons had recorded small variations in this radiation. The cosmic microwaves, it was discovered, were slightly redshifted on one side of the sky and slightly blue-shifted on the other. There was no way to avoid drawing the conclusion that the Milky Way did indeed have a peculiar velocity. In fact, the data were good enough that astronomers were able to deduce that the entire Local Group was moving through space at a velocity of about 600 kilometers per second.

Paradoxically, this result showed that the critics of the work of Rubin and Ford had been right, in a way. The measurements of the variations in the microwave background indicated that the Milky Way was moving in a direction almost exactly opposite to that found by the two Carnegie astronomers. Rubin and Ford thus found themselves in the unusual position of having their ideas vindicated at the same time that their results were found to be in error, because the results derived from measurements of the microwave background had to take precedence. Since the microwaves had been emitted in the big bang, they provided a frame of reference for the universe as a whole.

Astronomers soon concluded that the peculiar motion of the Local Group must be caused by the gravitational attraction of a concentration of mass that lay millions of light-years away, and could have no other cause. Although galaxies could presumably be set into motion by forces other than gravity—an exploding cosmic string could presumably induce a peculiar motion, for example—such events must have happened billions of years ago, and we would not expect that the resulting peculiar motion should persist today. The only logical explanation for the motion of the Local Group was that a Great Attractor was exerting a gravitational pull on the galaxies that were its constituents.

Astronomers weren't sure exactly how far away this hypothetical Great Attractor had to be. At least astronomical observations had revealed no great concentrations of mass in the area of the sky in which it was supposedly situated. On the other hand, it was a simple matter to calculate how massive it would have to be for any given distance. Newton's law of gravitation implied that the combined gravitational pull of several hundred extra galaxies could produce the observed motion, if that concentration of galaxies were 30 million light-years away. If the mass concentration lay at a distance of 300 million light-years, then a mass equal to that of tens of thousands of galaxies would be required.

STREAMING MOTIONS

One might think that finding the Great Attractor would be an easy task indeed, that all astronomers would have to do would be to determine which way the Local Group was moving, and to point their telescopes in that direction. Unfortunately, matters are not that simple. While it is possible to examine astronomical photographs and to pick out clusters and superclusters of galaxies, it is somewhat more difficult to guess how much mass they contain. Furthermore, measuring the motion of the Local Group alone does not tell astronomers the exact direction in which the Great Attractor lies. In fact, scientists would not expect the two directions to be the same, since the motion of the Local Group is also affected by the gravitational pull of a cluster of galaxies in the constellation Virgo.

In order to determine the location of the Great Attractor, we must first measure the motions of other groups of galaxies. If this is done, and if some sort of collective motion can be detected, then we have a chance of determining where the Great Attractor might be. In other words, the motion of one galaxy or group of galaxies means little, but if hundreds of galaxies are found to be moving toward the same point, then these data have a great deal of significance.

In 1987, a group of astrophysicists, who soon became

known as the Seven Samurai,* completed a five-year study of the distances and peculiar motions of about four hundred galaxies. The galaxies that they chose for inclusion in their survey were bright, elliptical galaxies distributed more or less uniformly in different directions in the sky. By centering their attention on this one type of especially bright galaxy, they hoped to avoid introducing bias in their data.

The study revealed that the motion of the Local Group was no small-scale effect. On the contrary, a concerted motion could be observed. According to the Seven Samurai, an enormous volume of the local universe, which included at least two superclusters of galaxies, exhibited a high-velocity streaming motion toward the (as-yet-undiscovered) Great Attractor. The Local Group, the cluster of galaxies in Virgo, and two superclusters in the Hydra-Centaurus and Pavo-Indus regions were all caught in the gravitational grip of some huge mass.

As the data were analyzed further, the outlines of the picture became clear. All the galaxies in our region of the universe were caught up in a streaming motion toward an attractor that had a mass at least 5×10^{16} times greater than that of the sun, and equal to that of tens of thousands of galaxies, and which was situated at least 400 million light-years from the Milky Way. The velocity of the streaming motion was about 600 kilometers per second in the vicinity of our galaxy. In places close to the Great Attractor, it rose to 1,000 kilometers per second or more.

It is now generally agreed that a Great Attractor exists, but astronomers are still not sure about its precise location. Some think that it is a giant supercluster of galaxies that—by a stroke of bad luck—is hidden from view by the dust in

---

* The Seven Samurai were David Burstein of Arizona State; Roger Davies of Kitt Peak National Observatory; Alan Dressler of Mount Wilson and Las Campanas Observatories; Sandra Faber of UC Santa Cruz; Donald Lynden-Bell of the Institute of Astronomy, Cambridge, U.K.; Roberto Terlevich of Royal Greenwich Observatory; and Gary Wegner of Dartmouth College.

the disk of the Milky Way, though there are other possible interpretations of the data. For example, some scientists think that the observed motions could be caused, not by a single Great Attractor, but by a number of smaller clusters of galaxies. Furthermore, the voids that exist in the universe may also play a role, since a void would create an absence of gravitational attraction that could cause galaxies to stream in the opposite direction, or at least contribute to their motion.

The Great Attractor could be a loop of cosmic string. Physicists Yehuda Hoffman and Wojciech Zurek of the Los Alamos National Laboratory have suggested that a loop some 330,000 light-years in diameter and having a mass 10^{13} times greater than that of our sun could produce the observed effects. However, other scientists are not so sure. For example, in a paper published in the British journal *Nature* in 1987, astrophysicists Adrian Melott of the University of Kansas and Robert Scherrer of the Harvard-Smithsonian Center for Astrophysics argued that cosmic strings could neither give large-scale streaming motions nor reproduce observed cluster–cluster correlations (a quantity that measures the clustering of the clustering of galaxies).

The Great Attractor could be a conglomeration of dark matter. There are, however, certain problems associated with this hypothesis. The most significant one is that the very existence of streaming motions seems to be inconsistent with cold-dark-matter theories. Calculations indicate that, if it is true that cold dark matter constitutes the greater part of the mass in the universe, and provided seeds for galaxy formation, then it must be so uniformly distributed throughout the universe that streaming motions of the observed magnitude would be impossible.

The existence of streaming motions does seem to be consistent with hot-dark-matter theories. But as we have seen, these theories have serious problems of their own, since they imply that the concentrations of matter that became clusters of galaxies formed before the galaxies themselves came into being, while the opposite seems to be the case.

Finally, unanswered questions about the streaming motion remain. For example, no one really has any idea whether the Great Attractor is stationary with respect to the microwave background, or whether it too is in motion. For that matter, there is still a certain amount of controversy regarding the size and significance of the streaming motion itself. This is a result of the fact that, if we want to determine the magnitude of the peculiar motions of the galaxies, we must know how far away they are from the earth. As we have seen, this measurement is notoriously difficult to make.*

THE FARTHEST THINGS IN THE UNIVERSE

The astronomers who study distant galaxies and other astronomical objects rarely speak of those objects' distance from the earth. If they did, they would only become embroiled in controversy. There is too much disagreement about distances in the universe.

Fortunately there is another way in which the locations of distant objects can be described: in terms of their redshifts. If an object has a redshift close to 0, this means that the light it emits is shifted by a very small amount, and thus that it must be relatively near the earth. If an object has a redshift of 1, this means it is moving away from the earth at so great a velocity that the wavelengths of light it emits are stretched by a factor of 100 percent. In other words, they are twice as long. A redshift of 1 corresponds to quite a large distance, incidentally. A simple calculation indicates that light that has been stretched by this amount must have been emitted when the universe was about half as old as it is now. If we

* As this book was being edited, several members of the Seven Samurai announced that they had pinned down the location of the Great Attractor more precisely. They found that its center lay about 150 million light-years from the Milky Way, and that it stretched some 300 million light-years across the sky.

assume that the universe is 15 billion years old, then a galaxy with a redshift of 1 would be about 7 billion light-years away.

As we look farther and farther out into space (and, accordingly, farther and farther back in time), redshifts increase rapidly. If we could see all the way back to the beginning of the big bang, redshifts would become infinite.

The largest observed redshifts are far from infinite. Until recently, the most distant object known was a quasar with a redshift of 3.78, which was discovered in 1982. This quasar has a velocity of recession greater than 90 percent of the speed of light. It is so far away that the light emitted by it that falls on the earth must have been emitted when the universe was only about 3 billion years old.

Quasars are bright objects that are thought to be the luminous cores of young galaxies. Because they produce such prodigious quantities of light, they can be seen at distances where other objects (ordinary galaxies, for example) would be invisible. Quasars are generally found at redshifts ranging from around 1 to around 3. At a redshift of about 2.5, their numbers begin to taper off, approaching 0 at a redshift of about 3.5.

Since modern telescopes are capable of seeing quasars out to redshifts of about 5, astronomers long believed that few or none would be found beyond the redshift 3.5 "limit." And of course, the discovery of a quasar with redshift 3.78 did nothing to disabuse them of that notion, since the difference between 3.5 and 3.78 isn't that great. Then, between August 1986 and September 1987, astronomers suddenly discovered seven new quasars with redshifts greater than 4. One of them, found in September 1987 by Stephen Warren, Paul Hewett, and Michael Irwin of Cambridge University, had a redshift of 4.43. Only a few weeks later, Mark Dickinson and Patrick McCarthy, two graduate students at the University of California at Berkeley, discovered a quasar with a redshift of 4.4.

When they made these discoveries, the Cambridge and

Berkeley astronomers were looking back to an early time indeed. A redshift of 4.4 (or 4.43) corresponds to a time less than 2 billion years after the big bang. They were not the only ones who were discovering objects that lay at the edge of the observable universe. Other astronomers were discovering evidence of normal galaxies almost as far away.

In 1983, J. Anthony Tyson of AT & T Bell Laboratories in Murray Hill, New Jersey, and Patrick Sweitzer, now at the Space Telescope Science Institute, had set out to study objects at the very limit of the observable universe. Using long photographic exposures and extensive image processing, they pushed observing techniques to new limits. Making observations with the 4-meter telescope at Cerro Tololo Interamerican Observatory in Chile, they selected sections of the sky relatively free of bright stars and galaxies in order to secure an uncluttered view of the deep universe.

By 1988, Tyson and Sweitzer had completed their survey, and had found some twenty-five thousand bright, fuzzy objects that looked very blue, and which had extremely high redshifts.* The redshifts of the brighter objects ranged from about 0.7 to 3, which meant that most of them lay at distances where only quasars were ordinarily observed.

Since the "blue fuzzies" were found at such high redshifts, Tyson and Sweitzer concluded that they must be newborn galaxies. Naturally the two astronomers couldn't be absolutely certain about this. The fuzzy objects were so far away that it was impossible to make out any details in their structure. However, the conclusion that they were galaxies seemed to be the only reasonable one.

According to Tyson, the discovery provided new information on the details of galaxy formation and evolution. Since the number of blue objects fell off rapidly at redshifts greater than 3, it appeared that one could conclude that galaxy

---

* It should be recalled that the redshifting of light does not make an object look red. These objects were blue because ultraviolet rays had been redshifted to the blue end of the visible spectrum.

formation most likely began at a redshift of about 4, and that star formation continued down to a redshift of about 1.

These findings—that quasars exist at redshifts of 4.4 and greater, and that the earliest galaxies began to form disks of stars at redshifts of about 4—present problems for the cold-dark-matter theory of galaxy formation. In the cold-dark-matter model, a certain amount of time is required to pass before star formation in galaxies can begin. According to the theory, the first concentrations of mass to come together would be the size of dwarf galaxies. Larger galaxies would begin to form only at a later epoch. Star formation could begin only some time after gravity had collected the primordial hydrogen and helium gas into galaxy-size objects. Calculations of the time required for all this to take place yield results just barely consistent with observations.

In other words, the existence of galaxies and quasars at these redshifts creates difficulties for the theory, but doesn't quite disprove it. However, if more distant objects were discovered, quasars at redshifts greater than 5, for example, the cold-dark-matter theory of galaxy formation would have to be discarded. Although, in many respects, it has been the most successful theory proposed so far, astronomers and cosmologists would have to look for an alternative theory that would predict that galaxies would form more rapidly.

CONFLICTING RESULTS

In this chapter, and in the previous one, I have made numerous references to the existence of dark matter in the universe. As I have explained, if this dark matter really exists, most of it must be nonbaryonic. That is, it must be composed either of neutrinos that possess a small but finite mass, or of as-yet-undiscovered objects (such as cosmic strings) or particles (such as WIMPs).

However, we should not forget that there is really only one reason for believing that this nonbaryonic dark matter exists. This is the prediction of the inflationary universe theories that the matter density of the universe must equal the critical density. If there was no inflationary expansion, then it would not be necessary to invoke the presence of hot or cold dark matter, cosmic strings, or balls of wall. If the inflationary universe paradigm were incorrect, then the matter density could easily be, say, a tenth of the critical value. In that case the dark matter in the halos of galaxies could easily turn out to be baryonic. The unseen mass could consist of dim stars or Jupiter-size objects, for example.

Consequently, it is important to consider the question of whether there really is any observational evidence that would force us to conclude that the inflationary expansion really took place. As we have seen, the inflationary theory is quite plausible, and it explains a great deal. However, this by itself should not be enough.

At present, the available evidence seems somewhat contradictory. For example, it is possible to estimate the total mass present in galaxies by observing their motions. Calculations show that the mass present in galactic clusters is between 10 and 30 percent of the quantity required to close the universe.

There could be dark matter in the spaces between clusters, enough to bring the total mass density up to the critical value. If this is the case, then dark matter does not cluster in the same way galaxies do. If it did, its presence would affect the galaxies' motions.

In other words, if nonbaryonic dark matter exists, it must be more or less evenly distributed throughout the universe. The dark matter itself could, in this case, be compared to oceans, and the concentrations of mass in galaxies to islands projecting slightly above sea level.

Calculations indicate that if nonbaryonic dark matter is distributed in this way, it ought to be cold dark matter. Neutrinos, on the other hand, would cluster in a different manner. However, as we have seen, the cold-dark-matter theory has begun to experience difficulties. No one can say that it is incorrect, but if it is correct, there are serious problems to be solved.

Other kinds of evidence also lead to inconclusive results. If astronomers could accurately determine the age of the universe, they might obtain some evidence to indicate whether the matter density was close to the critical value or not, since the quantity of matter present in the universe is related to its age. The more matter there is, the faster the expansion slows down. A critical-density universe would have been expanding faster in the past than one that contained less matter. This implies that a critical-density universe is younger; a faster early expansion means that it would take less time to reach its present state.

Calculations show that a critical-density universe would be about two-thirds as old as one that contained a much lower density of matter (such as the 10 to 30 percent cited above). Thus if astronomers knew exactly how rapidly the universe was expanding right now, they could compute the age that a critical-density universe could have. This theoretical value could then be compared to observations of various kinds.

Unfortunately, scientists don't know exactly how rapidly the expansion is proceeding. The uncertainties about the distances among the galaxies make this figure uncertain by more than a factor of two. As a result, all we can say is that a critical-density universe would have to be somewhere between 7 billion and 16 billion years old, while a universe with a much lower density (that is, one that contained only baryonic matter) could be between 10 billion and 25 billion years old.

Astronomers have determined that the oldest stars that can be observed are about 15 billion years old. This figure is just barely consistent with the 16-billion-year age of a critical-density universe, if we assume that star formation began when the universe was a billion years old (a short time by cosmological standards). If measurements of the expansion rate were refined and the 16-billion-year upper limit lowered, then there would be a conflict. We would know either that there was something wrong with the estimates of stellar ages, or that an important prediction of the inflationary universe theories had been contradicted.

In 1988, University of Hawaii astronomer Brent Tully presented results which seemed to indicate that there was such a conflict. Tully's work, which was based on computer models and the known recession velocities of distant galaxies, seemed to imply that the expansion of the universe was more rapid than the majority of astronomers had believed. According to Tully, the gravitational drag of groups of galaxies near the Milky Way had introduced errors into many previous estimates.

If Tully's results are correct, a critical-density universe could only be 7 to 10 billion years old. A universe with a much lower matter density, on the other hand, could be 50 percent older than that. Thus, if we assume that nonbaryonic dark matter does not exist, the discrepancy implied by Tully's results is much smaller.

MEASURING THE CURVATURE OF SPACE

We should not hastily conclude that there was no inflationary expansion. On the other hand, a team of Princeton University astronomers has obtained results which seem to indicate that the prediction of the inflationary universe theories, that the matter density lies near the critical value, is indeed correct.

The two astronomers, Edwin Loh and Earl Spiller, did not set out to measure the matter density directly. Obviously, this would be impossible, since scientists are not sure what the nonbaryonic matter in the universe (if it exists) is made of, nor how it is distributed. Trying to measure it directly would be a hopeless task. However, if it is present, this should have observable consequences; one should be able to detect it indirectly.

According to Einstein's general theory of relativity, the matter density of the universe is related to the average curvature of space, which is in turn related to the numbers of galaxies that will be seen at certain distances. A simple analogy should show why this should be the case. Imagine that you are standing at the earth's North Pole and looking out along an imaginary circular plane that stretches 12,500 miles in every direction. The area of this circle will be about 490 million square miles. Compare this to the surface of the earth, which is only about 200 million square miles, even though the South Pole is the same distance away—12,500 miles—as the edge of the circle. In other words, if the two-dimensional surface of the circle is curved to fit the curved surface of the earth, its area will be much diminished.

Curvature affects the three-dimensional space of the universe in a similar way. The greater the curvature, the smaller the number of galaxies that will fit into it. Here, we make the assumption that galaxies are equally spaced. Of course galaxies are not equally spaced. They are grouped in clusters, and the spacings between clusters are not constant either. This really presents no problem, however, since we can compute average spacings. If the volume of space studied is large enough, the averaging process will eliminate any errors that might be caused by irregularities.

In their study, completed in 1986, Loh and Spiller counted the number of galaxies that could be observed in selected volumes of space. Then they divided these numbers by the volume of space in each region to obtain averages that would allow them to compute average spatial curvature. Finally, they used this figure to find the average mass density.

When the computations were completed, they found that the universe had a mass density somewhere between 60 and 120 percent of the critical value. Though various kinds of observational uncertainties made it impossible to be more precise, this result did seem to confirm the prediction of the inflationary universe theories that the mass density was at the critical value. As we have seen, a universe that contained only baryonic matter, in which no inflationary expansion had presumably taken place, would have a density between 0.1 and 0.3, figures that fall outside the range found by Loh and Spiller.

WAS THERE AN INFLATIONARY EXPANSION?

Most astronomers and astrophysicists would like to believe that there was. The theory explains so much that they would be reluctant to give it up. If the theory turned out to be false, new ways would have to be found to solve the problems associated with the original big bang theory, and this would not be easy.

Furthermore, recent work has made it more difficult to believe that an inflationary expansion did not take place. It now appears that we need make no special assumptions about the early universe in order to reach the conclusion that there was such an expansion. It appears that any of a number of different kinds of fields could produce it. Guth's original theory postulated that the expansion was driven by phase transitions in the fields associated with the Higgs particle, but it is now apparent that, if it turns out that the Higgs particle does not exist, this will hardly be fatal to the inflationary paradigm.

In the end, theory must submit itself to the scrutiny of experiment and observation, and here the situation is somewhat more ambiguous. Tully's results, for example, taken in conjunction with determinations of the ages of certain stars, would seem to indicate that there was no inflationary expansion, or at least that there is something wrong with the prediction that the universe should have the critical density.

Perhaps the Loh and Spiller results must be considered slightly more convincing, since their work was an attempt to measure the matter density directly. However, there are reasons for being cautious about their findings also. Any one of a number of different factors could have introduced errors into their results. For example, galactic evolution could have an effect. As astronomers look out into space, they also look back into time, but no one is really sure whether galaxies were brighter or dimmer than they are now, or whether they had approximately the same luminosity. They may become dimmer as stars age and die, or they may become brighter as gravitational attraction causes larger galaxies to cannibalize smaller ones. If either effect is important, the final figure for the mass density of the universe could be in error.

Did an inflationary expansion take place? At the moment, we would have to lean toward the conclusion that it did. However, a certain amount of caution is necessary. There are still unresolved problems concerning dark matter and the age of the universe. The evidence that supports the inflationary universe paradigm are hardly very convincing. The reasons for believing that there was such an expansion are largely theoretical.

LATE NEWS FLASHES

While this book was being edited, a series of astonishing new discoveries were reported. However, these discoveries do not seem to have cleared up any of the outstanding problems in cosmology. If anything, the situation is even more confusing than it was before.

The problem is that no one really understands how the new findings can possibly be consistent. On one hand, it has been found that the big bang was a very, very smooth explosion. Satellite measurements of the cosmic background radiation carried out in late 1989 revealed no trace of any lumpiness in the early universe that might later have evolved into galaxies and clusters of galaxies. Other findings, announced in late 1989 and in early 1990, indicate that the present-day universe is very lumpy indeed, that it contains huge structures whose presence had not previously been suspected.

On November 18, 1989, NASA's Cosmic Background Explorer (COBE) satellite was launched. Measurements of the cosmic background radiation carried out by the satellite allowed scientists to look back to within a year after the big bang; they were able to see farther back in time than had ever been possible before. The measurements that they obtained revealed only a perfect smoothness. There were no bright spots in the radiation, or variations of any other kind. This seemed to indicate that the matter density of the early universe was also perfectly even. After all, any lumpiness in the distribution of matter would have produced a corresponding lumpiness in the radiation that was emitted.

But the day before the COBE spacecraft had been launched, Margaret J. Geller and John P. Huchra of the Harvard-Smithsonian Center for Astrophysics in Cambridge, Massachusetts, had announced their discovery of a "Great Wall," a huge concentration of galaxies that was located 200 to 300 million light-years from earth. The Great Wall, they found, was approximately 500 million light-years long, 200 million light-years wide, and 15 million light-years thick.

But this was only the beginning. At about the same time that Geller and Huchra published their results, two teams of astronomers in the United States and Great Britain were sharing data that they had been collecting over the previous seven years. The findings of the two teams were compared and, in early 1990, the astronomers reported that the Great Wall was only one of a very large number of massive clumps in the universe. Not only were there many concentrations of galaxies like it, these clumps appeared to be almost evenly spaced.

Depending upon the assumptions that were made about how rapidly the universe was expanding (as we have seen, there is still considerable disagreement on this matter), the clumps were 400 to 800 million light-years apart. So regular was their distribution that they gave a honeycombed appearance to the universe.

The existence of this kind of structure seemed to contradict the findings obtained by the COBE satellite. The existence of such structures seemed to imply, according to astronomer David C. Koo of the University of California at Santa Cruz,* that "an inherent roughness" had been imprinted on the universe a fraction of a second after the big bang. And yet the COBE measurements had revealed no roughness at all.

* Koo was one of the authors of the article in the British journal *Nature* in which these findings were reported. The other authors were Thomas Broadhurst and Richard Ellis of the University of Durham in England and Richard Kron and Jeffrey Munn of the University of Chicago.

III

BEYOND THE FRONTIERS: THE BOUNDARIES OF SCIENCE

8

Superstrings: Twenty-First-Century Physics or Medieval Theology?

SUPERSTRING THEORY has been described by some as a kind of twenty-first-century physics that was discovered by accident during the twentieth century. "No one invented it on purpose," says Princeton University physicist Edward Witten, "it was invented in a lucky accident. By rights, twentieth-century physicists shouldn't have had the privilege of studying this theory."

Other scientists have compared the discovery of superstring theory to the discovery of relativity and quantum theory early in this century. Some have expressed the conviction that it will prove to be the long-sought "theory of everything," a theory that would explain all the interactions of all the fundamental particles, a theory from which all of the other laws of physics could be derived.

As we saw in chapters 1 and 2, the standard model of particle interactions is a perfectly adequate set of theories in the sense that there is no experimental data that would contradict it. However, as I have also pointed out, theoretical physicists have never been really satisfied with this model. They would like to have a theory that would explain why there are three (or possibly four) families of quarks and leptons, why the individual quarks and leptons have the masses they do, why positive and negative electric charges come in particular sizes, why there are four forces, and why these forces have such widely varying strengths.

A theory that explained these, and certain other, properties of the fundamental particles and forces would not be a "theory of everything" in a literal sense. After all, if a theory of everything were found, there would still be a lot of work for physicists to do. However, they would no longer be seeking the basic laws of nature upon which everything else was built.

Some scientists have questioned whether such fundamental laws exist. In their view, scientists will never find a theory of everything because there can be no such thing. University of Texas theoretical physicist John Archibald Wheeler, for example, has expressed this view in a letter to me. "I can't agree that there's any magic *equation!*"

Others have expressed skepticism about superstring theory in particular. For example, Nobel prize–winning physicist Sheldon Glashow and his Harvard University colleague Paul Ginsparg have likened superstring theory to medieval theology. "Contemplation of superstrings," they write, "may evolve into an activity . . . to be conducted at schools of divinity by future equivalents of medieval theologians. For the first time since the Dark Ages, we can see how our noble search may end, with faith replacing science once again." The late Richard Feynman, another Nobel laureate, once expressed a similar opinion in his typically brash manner.

According to Feynman, superstring theories were "nonsense."

There have been numerous times in the history of science during which new theories were greeted with skepticism, but, as far as I know, there has never been a time when a new theory produced such excitement among its adherents while provoking such contempt from its opponents.* It is obvious that, whether superstring theory turns out to be true or false, it must be something very remarkable. There have not been many scientific theories believed by some to be capable of explaining "everything," which were simultaneously compared to medieval theology by others.

POINT PARTICLES

In order to see why the adherents of superstring theory should consider these theories (as we shall see, there are several) to be so exciting, it is necessary to understand something about the problems associated with conventional theories of particle interactions. These problems plague even such extremely successful and well-confirmed theories as QED (the reader will recall that QED is the theory that explains the forces that cause electrically charged particles to attract and repel one another).

The problems come about because these theories treat elementary particles as though they were mathematical points. Now, a dimensionless point is a mathematical abstraction. There is no particular reason to believe that an elementary particle should have such a character, and there are good reasons for thinking that it doesn't. Nevertheless,

---

* To be sure, there were controversies about relativity and quantum mechanics when those theories were introduced, but I don't think that these had the same intensity. Relativity, for example, was accepted relatively (pun only half intended) quickly by the scientific community. Physicists weren't as much taken aback by its predictions as the lay public seems to have been.

until the advent of superstring theory, physicists persisted in constructing theories that viewed elementary particles as though they were dimensionless. They did this because it appeared that they had no choice.

In order to see why this should be so, I will consider the case of the electron. I might as well begin by assuming that an electron is a tiny sphere. If it had a different shape, this would not affect the following arguments. Once it is assumed that the electron has a spherical shape, a question arises: Can the electron be deformed, or is it perfectly rigid? When physicists considered this question, they quickly discovered that either answer landed them in difficulties.

In the everyday world, there are no perfectly rigid objects. Though a golf ball, for example, may feel perfectly hard and rigid to the touch, it is not so in reality. When a golf ball is hit with a club, the entire ball does not begin moving all at once. First the ball becomes deformed at the point of impact; in other words, the part of the ball struck by the club begins moving first. The rest of the ball is set into motion only as the resulting shock wave travels from one side of it to the other. To the eye, it may seem that the ball is set into motion instantaneously, but a high-speed camera would reveal that something much more complicated is going on.

In fact, there can be no such thing as a perfectly rigid body in nature. If a golf ball were that rigid, and the entire ball began moving at once, then the shock wave would have to travel through the ball at an infinite velocity. This is forbidden by Einstein's special theory of relativity, which states that no signal or causal influence can travel at a velocity greater than that of light. Thus it appears that if we accept the strictures of relativity—which is one of the best-confirmed theories in physics—then we must conclude that neither a golf ball nor our hypothetical spherical electron can be perfectly rigid.

If an electron is not rigid, then it must be possible to deform it in the manner that a golf ball is deformed. Unfortunately, making this assumption creates serious problems

too. If electrons could be deformed, this would create observable effects that would show up in experiments, but no
such effects are seen. Furthermore, if we could stretch and
bend an electron, then there would be no reason why it
should not be possible to pull the electron apart, but fragments of electrons are not seen in nature.

Viewing the electron as a dimensionless point also creates
difficulties, but these turn out to be problems that can be
solved, or at least evaded. For example, the assumption
that an electron is a mathematical point leads to the conclusion that it must have infinite mass. However, there exists a procedure for sweeping this unpleasant—but not
unexpected—result under the rug. This procedure is called
renormalization.

A point electron would have infinite mass because the
electron is a charged particle. In order to see why this would
be the case, imagine that an electron is broken up into
several different pieces. Now, the laws of electromagnetism
tell us that like charges will repel one another, while unlike
charges will attract. Thus there will be a repulsive force
between the negative charges of the various different pieces
of the electron. Furthermore, the more closely these pieces
are brought together, the stronger these forces will become.
At zero distance, when the different pieces of the electron
are compressed together into a point, these forces will become infinite. Obviously, it would take an infinite amount of
energy to overcome an infinite repulsive force, but if an
electron had an infinite energy, it would also have an infinite
mass. This is a consequence of Einstein's equation $E = mc^2$.

Obviously, the electrons encountered in nature have neither infinite energies nor infinite mass. In fact, the mass of
the electron has been determined to great accuracy, and it
turns out to be small indeed. It is 0.511 MeV, or about 9 ×
10^{-28} grams.

In QED, the assumption is nevertheless made that electrons are dimensionless points. At first glance, it seems
astonishing that a theory based on such an absurd postulate

could turn out to be so successful. However, the theory is "saved" by the fact that no one has ever seen a bare electron. Quantum mechanics tells us that there is no such thing as nothing, that "empty" space is never really empty. Thus an electron must always be surrounded by clouds of virtual particles, which shield it and prevent us from seeing its infinite mass.

Renormalization is a mathematical technique that was developed to deal with infinite masses, and with other infinite quantities that crop up in QED, and render them innocuous. When this technique is applied, an infinite energy associated with the cloud of virtual particles is subtracted from the infinite self-energy of the electron, and one obtains a finite result.

When infinities are encountered in a scientific theory, this is generally a sign that something has gone wrong, that the theory harbors contradictions of some sort, or that there is something wrong with the initial assumptions. If the postulates of a theory in which infinities are encountered cannot be changed so that the infinite quantities disappear, then the theory will generally have to be discarded. We would not expect, therefore, that a technique such as renormalization, which is mathematically questionable to begin with, should produce acceptable results.

Surprisingly, the technique gives more than acceptable results. When QED is renormalized, it produces predictions that can be confirmed experimentally to a degree of accuracy that is rare in physics. The renormalized version of QED is accurate down to dimensions much smaller than an atomic nucleus, and its predictions have been verified to an accuracy of better than one part in a billion.

Both of the theories that make up the standard model can be renormalized also. The procedure can be applied to the electroweak theory (which incorporates QED) and to quantum chromodynamics (QCD). Furthermore, QCD and the electroweak theory can be combined in GUTs which, as we have seen, represent an attempt to unify three of the four forces of nature: the strong and weak forces and electromagnetism.

To be sure, several different GUTs exist and no one knows which of them, if any, is most likely to be correct. Furthermore, the GUTs make predictions that have not yet been verified by experiment. However, it appears that making the seemingly unrealistic assumption that the world is made of point particles produces results that are much better than one would have a right to expect.

BUT WHAT ABOUT GRAVITY?

If one or another of the GUTs were eventually verified, this would represent a great advance. However, even in that case, theoretical physicists would remain unsatisfied. The GUTs do not seem capable of explaining "everything." There would still be a number of parameters—particle masses and so on—that would not be specified by the theory, but which would have to be determined by experiment. Furthermore, scientists would still have to work with two interactions, the combined strong-weak-electromagnetic force, and gravity. In an ideal world, all four forces could be understood as different aspects of a single superforce.

Unfortunately, it is difficult to combine gravity with the other three forces. Specifically, it has not yet proved possible to develop a quantum theory of gravity. When physicists have attempted to understand gravity as a force transmitted by the hypothetical particles known as gravitons, theoretical nonsense has been the result.

It has proved to be impossible to carry out the renormalization procedure in the case of gravity. Like the other quantum field theories—QED, electroweak theory, QCD, and the GUTs—quantum gravity theory produced infinities, but these infinities were much worse than those encountered in other theories. There appeared to be no way to get rid of them.

It wasn't difficult to understand the source of the difficulty. Gravity is a more complicated force than the other three. If general relativity is correct—and there is ample experimental evidence that seems to indicate that it is—then it is necessary to conclude that the creation of gravitational energy creates additional force. In other words, gravity gravitates; a gravitational field attracts itself.

In the language of quantum field theory, this means that gravitons must interact with one another in ways that other force-carrying particles do not. When photons act as carriers of the electromagnetic force, for example, they blithely ignore one another's presence. Gravitons, on the other hand, apparently do not; they interact with one another as well as with the gravitating bodies that emit and absorb them.

Furthermore, there seems to be no way around the mathematical difficulties that result. There is no "super-renormalization" procedure that would clear up the problem. It appears that one must conclude that two of the most successful theories in the history of physics, quantum mechanics and general relativity, are inconsistent. Though physicists are sure that both are correct, they have no idea how the two can be combined.

A THEORETICAL BACKWATER

While the physicists who worked in the theoretical mainstream struggled unsuccessfully with the problems of the unification of all four forces, and with quantum gravity, a small number of scientists working in a theoretical backwater pursued ideas that most physicists considered to be quite unpromising. Furthermore, as these ideas were subjected to more intense scrutiny, it quickly began to appear that they are not just unpromising, they were downright absurd. For example, some of the theories being developed seemed to imply that space might have, not three, but as many as twenty-five dimensions.

In 1968, CERN physicist Gabriele Veneziano discovered a mathematical formula that seemed to describe certain properties of hadrons (hadrons are particles that feel the strong force; "hadron" is a collective term that applies to both baryons and mesons). Although Veneziano's model was very

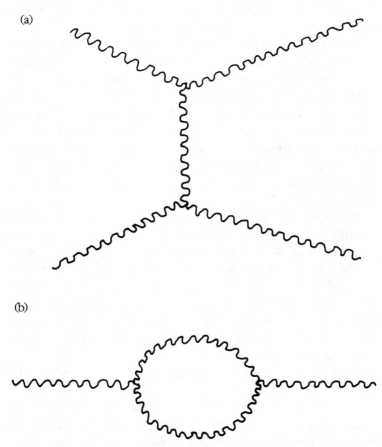

Gravity gravitates. In (a), one graviton emits a second graviton, which is absorbed by a third. In (b), a single graviton decays into two, which merge to form one graviton again. Interactions of this type, which other kinds of particles do not exhibit, complicate matters considerably, and have made it impossible, so far, to construct a quantum theory of gravity.

successful in some respects, it contained certain mathematical inconsistencies.

It soon became apparent that these inconsistencies could be eliminated. However, the cure, as the saying goes, appeared to be worse than the disease. In order to accomplish this task, it seemed to be necessary to formulate the theory, not in four,* but in twenty-six** dimensions. In other words, the theory only worked if there were an extra twenty-two dimensions that had never been observed.

When given the choice between entertaining a seemingly absurd idea or accepting mathematical inconsistencies, physicists always choose the former. After all, they are aware that science often discovers things that are contrary to common sense. They also know that the presence of a mathematical inconsistency is much worse because it will eventually lead to contradiction, and, naturally, they could place no trust in a theory that was likely to contradict itself at any moment. It is far better to force ourselves to believe the "impossible."

It wasn't obvious, however, how these extra dimensions should be interpreted, if indeed they really did exist. Only three spatial dimensions are observed in the everyday world. Furthermore, there were mathematical proofs which seemed to show that if space had more than three dimensions, then gravity could not have the form that is observed. For example, in a space of two, or four, or more, dimensions, the planets in the solar system could not move in stable orbits around the sun.

Nevertheless, the physicists pushed on undaunted, and studied Veneziano's theory more deeply in an attempt to force it to reveal its secrets. Finally, in 1970, the Japanese-American physicist Yoichiro Nambu showed that the mathematical formula Veneziano had found could be interpreted

---

* That is, three dimensions of space and one of time.

** In this case, there would be twenty-five spatial dimensions. Time would be the twenty-sixth.

in an intriguing way. This formula could be derived by making the assumption that hadrons were not point particles, but rather one-dimensional strings that vibrated in a twenty-six-dimensional spacetime.

Although Nambu's string theory (it was not yet a theory of superstrings) aroused some interest, it quickly fell into disrepute. Not only did it fail to explain why the extra dimensions were not observed, it seemed to possess inconsistencies of its own. Though Nambu's theory purported to be a theory of hadrons in general, it was soon discovered that a twenty-six-dimensional string theory could only describe bosons, the particles associated with forces. It could not be applied to protons or neutrons, or to other particles of matter, which were fermions.

Whatever interest there had been in string theory quickly subsided. The idea that hadrons were made, not of strings, but of quarks, seemed much more promising. Theoretical physicists turned their interest to the development of QCD, and string theory became a theoretical backwater.

Even when it was shown that the behavior of fermions could be described by a ten-dimensional theory, interest in strings did not revive. The meaning of the extra dimensions was still unexplained, and there were other difficulties as well. For example, the theory seemed to require the existence of spin-1 and spin-2 bosons that looked more like the photon and the graviton than the fermions the theory was trying to describe. As a result, most physicists soon became convinced that the concept of particles as strings was yet another idea that had briefly seemed intriguing, but which had proved to be unsuccessful in the end.

SUPERSTRINGS AND GRAVITY

A few scientists did continue to work on string theory. In 1974, the French physicist Joel Scherk and John H. Schwarz of the California Institute of Technology showed that the presence of these extra particles in string theory was a virtue, not a defect. If one conceived of strings as tiny objects about 10^{-33} centimeters in length, then the theory could be used to unify gravity with the other three forces. Furthermore, the force of gravity predicted by the theory would have the right strength. The presence of spin-2 gravitons in the theory was apparently not an anomaly after all.

However, interest in string theory did not increase when this finding was published. On the contrary, it declined further. At the time, it was becoming obvious that the theories that made up the standard model were capable of explaining all the experimental data that were then available to physicists. There seemed to be no need to investigate new, and admittedly esoteric, ideas. By the end of the 1970s, the concept of particles as strings had been virtually forgotten.

Then, in 1984, the theoretical situation changed suddenly. In that year Schwarz and Michael Green of Queen Mary College in London showed that a particular string theory that incorporated the concept of supersymmetry was free of certain mathematical inconsistencies, known as anomalies, that had plagued string theory from the beginning.

Unlike the original theory of Nambu, the theory of Schwarz and Green was a superstring theory. The name, in fact, is nothing more than shorthand for "supersymmetric string." Supersymmetry is a concept that I discussed briefly in chapter 5, where I pointed out that it is based on the idea that there are not two kinds of particles in nature, but only one. Supersymmetric theories put fermions and bosons on the same footing, and imply that every fermion has a boson "partner."

The concept of supersymmetry has a great deal of appeal because it makes any theory into which it is incorporated seem simpler, but it is more than a beautiful mathematical concept. Introducing supersymmetry turns out to be a way of unifying the forces. When the consequences of supersymmetry are worked out in detail, it is found that any theory that is supersymmetric will automatically include the force of gravity.*

Of course, not all supersymmetric theories work. In fact, when Schwarz and Green published their result in 1984, theoretical physicists were just beginning to conclude that another supersymmetric theory, known as supergravity, could not predict observed experimental facts. Supergravity, which was also a multidimensional theory (the most popular version had eleven spacetime dimensions) differed from superstring theory in that it conceived of particles as mathematical points.

When Schwarz and Green published their paper, the response was immediate. Theoretical physicists around the world rushed to familiarize themselves with the ideas of superstring theory. Within the space of a few years, superstrings had become the main focus of advanced theoretical research.

Obviously, this didn't happen only because Schwarz and Green had eliminated some mathematical anomalies. The resurgence of interest in superstrings can be attributed to a number of factors. One was growing dissatisfaction with the standard model. More and more physicists had begun to feel that it just didn't explain enough. Another factor was increasing interest in the idea of unification, coupled with the realization that the rival supergravity theories were probably not going to work.

ROLLING UP THE EXTRA DIMENSIONS

Yet another factor that spurred interest in superstrings was the rediscovery of some theoretical work that had been done

---

* Here I should add, for the benefit of those who are familiar with certain technical points, that only *local* supersymmetry implies gravity; the less stringent condition of *global* supersymmetry does not.

in the 1920s, when the Polish physicist Theodor Kaluza had shown that extra spatial dimensions could be interpreted as forces, and the Swedish physicist Oskar Klein had demonstrated that these extra dimensions could be rolled up, or compacted, to such a degree that their presence could never be directly perceived.

The concept of compactification is not as esoteric or as complicated as one might think. In a way, any one of us is capable of compacting a dimension at any time. Naturally, we can't roll up dimensions in the space around us. However, it is possible to pick up a sheet of paper, roll it into a cylinder, and then roll up the cylinder more and more tightly. As we do this, one of the dimensions of the two-dimensional sheet of paper becomes compacted, and the cylinder's diameter grows progressively smaller.

Naturally, there is a difference between a compacted sheet of paper and a compacted dimension of space. I doubt that it would be possible to roll up a sheet of paper to a diameter of much less than a centimeter or so. The extra dimensions of superstring theory, on the other hand, are rolled up to a size of about 10^{-33} centimeters; their size is approximately the same as that of a string.

Now, 10^{-33} centimeters is a quantity about 10^{20} times smaller than the diameter of an atomic nucleus, which measures about 10^{-13} centimeters across. It is obvious that neither superstrings nor compacted dimensions will ever be observed directly, since the energy required to probe matter to that depth is simply too great. Even if we could construct a particle accelerator as large as our solar system, the energy produced would be many orders of magnitude too small.

Theoretically, superstrings could be either open or closed. An open string would have free ends, while a string of the latter variety would form a closed loop. The strings in Nambu's original theory were open. Though both kinds are possible in modern theories, the idea that the superstrings are closed loops is generally considered to produce more promising results. Furthermore, though the original string

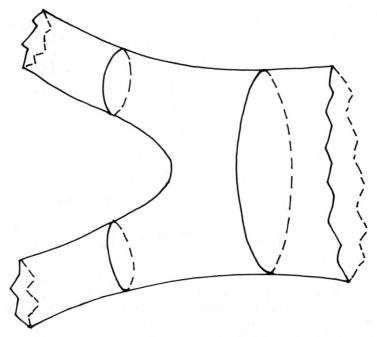

A *"pants diagram."* *Two closed strings may join and form a single string. In superstring theory, something like this happens when one particle "absorbs" another. One string has one closed loop where two existed before. Similarly, if the diagram were turned around, it would depict a situation in which a single particle (one loop of string) decayed into two.*

theory had twenty-six dimensions, all superstring theories nowadays are ten dimensional. It has been shown that they must have this dimensionality if they are to be consistent.

THE INFINITY PROBLEM AND OTHER TERRORS

If superstring theory is correct, then the fundamental constituents of matter are not point particles. On the contrary, they have a small but finite size. This fact has created hope that the infinities that plague quantum field theories might

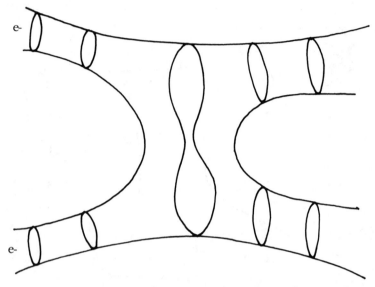

Interaction of two electrons. More complicated interactions are naturally possible. Here, two electrons interact with one another as they travel through space. Two loops of string briefly merge, and then separate again.

be absent from superstring theory, and that the mathematically questionable renormalization procedure might not be required.

Unfortunately, no one knows whether superstring theories contain infinities or not. A theory based on simple assumptions—and the idea that particles are basically vibrating strings is certainly simple—can become mathematically quite complicated when it is worked out in detail. Superstring theories are complicated indeed, so much so that no exact solutions for the mathematical equations associated with them have been found. Nor do theoretical physicists entertain any hopes that exact solutions will be obtained in the foreseeable future.

Those who work with superstring theory must therefore rely on a mathematical procedure of successive approximations known as perturbation theory. When perturbation the-

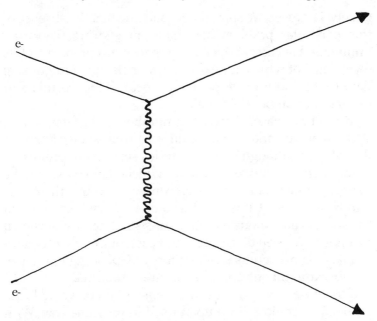

Here, the same process that is shown in the figure on page 25 is depicted in a conventional Feynman diagram. If the electrons are viewed as point particles, rather than loops of string, they do not "merge" at all (this would be impossible), but simply exchange a photon. No one yet knows which picture is more nearly correct; superstring theory is still very speculative.

ory is used, one proceeds step by step, making a first-order approximation, then a second-order approximation, and then (if the equations have not already become too complicated for this to be done) a third, and so on. So far, the approximate solutions obtained in superstring theory do not contain any infinities, but it does not necessarily follow that infinities will be absent at higher orders of approximation, and the number of higher orders is endless. In perturbation theory, we approach the exact solution more and more closely at every step, but never quite reach it. To obtain an exact solution, an infinite number of steps would be required.

Nevertheless, the fact that no infinities have turned up so far is considered to be promising. The situation is unlike that encountered in other theories, where they can pop up immediately. However, we should not conclude that the absence of this particular problem has made superstring theory an unqualified success. Superstring theory has problems of its own, some of which appear so formidable that it might take decades to find ways to deal with them, if indeed this ever turns out to be possible at all.

In the first place, there are a number of different superstring theories, and others could very well be discovered in the future. Although certain theories seem more promising than others, no one really knows which is most likely to be correct. In fact, depending upon how we count them, the number of superstring theories could be anywhere between six and several thousand. Six consistent ten-dimensional theories have been discovered, but each of these six theories can take on numerous different forms depending upon how the six extra dimensions of space are compacted.

If there were only one extra dimension, there would be no problem. It could roll up upon itself in only one way. With six, on the other hand, the possibilities are numerous; the six compacted dimensions can roll up in and around one another in numerous different configurations. Physicists have no idea which of the many different geometries that result corresponds to that which is most likely to be encountered in the real physical world.

For that matter, scientists do not understand why six dimensions should be compacted while the other four are not. The problem, incidentally, isn't one of understanding why six dimensions roll up; on the contrary, it is one of comprehending why our familiar four dimensions are not compacted too. And this is only the beginning. Like all other theories in physics, superstring theories are formulated in space and time. Space and time, after all, are the basic components of our world, or at least they have always seemed to be. No one knows how one would go about working out a theory that did not depend on them.

Yet many theoretical physicists have the nagging suspicion that, in the case of superstring theory, this procedure will turn out to be incorrect. They suspect that, in some sense, space and time are built out of the superstrings themselves. At present, they don't know how to deal with this problem. However, some scientists suspect that superstring theory will eventually change our ideas about what space and time are.

Finally, superstring theory also has problems of a more mundane nature. These are similar to problems we have encountered before in discussions of other theories. Since superstring theory is mathematically so complicated, physicists have been able to derive few specific predictions from it, and the few predictions that have been obtained are contradicted by the observed facts.

Superstrings are conceived as objects that vibrate in ten dimensions. Different levels of vibration correspond to different observed particles. In particular, the vibrations that have the lowest energy correspond to particles of zero mass. The next lowest energy levels produce particles with masses of about 10^{19} GeV, which is about 10^{19} times greater than the mass of a proton, or slightly less than a GeV—10^{19} GeV is approximately the mass of a dust particle.

Obviously, this result is not confirmed by experiment. A few particles, such as photons, and possibly gluons and neutrinos, have masses equal to zero, but the other particles do not. Those that do have masses cannot correspond to the 10^{19} GeV particles that the theory predicts. That quantity is many orders of magnitude too large.

Superstring theorists are not deterred. They point out that, as superstring theory is developed further, it is not unreasonable to expect that the theory will produce small corrections that will give the right masses after all. Nor are they bothered by the fact that no elementary particles with mass of 10^{19} GeV have ever been seen. Existing particle accelerators, after all, could no more produce particles with masses that large than they could be used to probe the structure of matter at distances of 10^{-33} centimeters.

On the other hand, the fact that the theory produces

predictions like this hardly leads to an ideal situation. If superstring theory is correct, if the natural mass scale for elementary particles turns out to be multiples of 10^{19} GeV, then one has to conclude that our entire macroscopic world, which is built out of protons and neutrons that weigh slightly less than a GeV, and of electrons that are even lighter, is the result of hyperfine corrections to a ten-dimensional theory.

BUT ARE THERE REALLY TEN DIMENSIONS?

It now appears that there may not be. Though it originally seemed that superstring theory had to be formulated in ten spacetime dimensions, the most recent results indicate that this might not be absolutely necessary. In fact, Edward Witten has worked out a way to formulate superstring theory in the usual four dimensions. In Witten's theory, the extra six dimensions are still present, but they do not have spatial character. However, it is not yet understood precisely what these six extra "things" in the four-dimensional theory really are. Apparently all that can be said is that the theory requires six additional variables of some sort.

A theory formulated in four dimensions is not necessarily different from one that requires ten. It is more likely that the two formulations are simply the same theory in two different guises. Naturally, there are some conceptual problems here, and these problems might not be cleared up until scientists are able to understand how superstring theory should logically be formulated and what space and time really are.

THEORY AND EXPERIMENT

If anything is obvious, it is the fact that the human mind is capable of constructing theoretical systems of the most far-reaching and fantastic character. Some of these theoretical fantasies are suspect indeed. We can find examples in almost every field of human endeavor, ranging from pseudo-sciences such as astrology to metaphysical excesses in the field of philosophy to aberrations within science itself. Nothing is more obvious than the fact that there is no idea so bizarre that it has not been believed by some human beings in some place at some time.

Scientists generally depend upon experiment to keep their theoretical flights of fancy in check. For example, nineteenth-century physicists believed that light could not be propagated through empty space, that some medium, called the ether, or luminiferous ether, was required to carry it. Scientists theorized about this nonexistent ether to the extent that their ideas became ridiculous. According to one leading physicist, for example, the luminiferous ether, which could be neither seen nor felt, had a density of thousands of tons per cubic millimeter, and was "squirming with the velocity of light." Such ideas were finally put to rest only when Einstein demonstrated that the idea of an ether was (in his word) "superfluous."

Since that time, numerous other fantastic ideas have been overturned (and sometimes confirmed) by experiment. When this has happened, it has more often than not led to further scientific advances, for when scientists discover that one of their most cherished theoretical ideas doesn't work, they must seek one that does.

Unfortunately, superstring theory has not yet made contact with experiment. Nor is it likely to do so during this century. The theory is too new, too dissimilar to previous theories, and mathematically too complicated. The fact that the quantities that physicists would like to use it to predict, such as the masses of observed particles, are at best the results of undiscovered hyperfine corrections, doesn't make matters any better.

As a result, there is a real danger that an entire generation of theoretical physicists could discover, a decade from now, or several decades from now, that they had been pursuing a theoretical chimera. It is indeed possible that superstring theory, as Feynman suggested, could turn out to be nothing but "nonsense."

Yet superstring theory continues to be the center of attention within the community of theoretical physicists. Virtually all the best minds within that community have become preoccupied with it. To be sure, some of them, like Glashow and Feynman, have rejected it, but the majority have not.

Superstring theory, after all, is what we sometimes like to call "the only game in town." Though scientists have explored every possible idea, no one has found any other plausible way of unifying the four forces within one theory, and, as we have seen, the unification of the forces is the theoretical physicist's Holy Grail. This goal may or may not be attained, but if it is reached, scientific understanding of the fundamental principles of nature would be so enriched that the benefits would be enormous.

Nobel Prize–winning physicist Steven Weinberg has expressed an idea that seems to be common among theoretical physicists nowadays. Weinberg does not claim to know whether or not superstring theory will ultimately prove to be correct. "Whether string theory will have been a good idea," he says, "depends on what comes out of it." But it would be "crazy" not to pursue the theory, he adds. He admits that, in the end, physicists may find that there are "insuperable obstacles to finding an interpretation that corresponds to physical reality." However, he adds, "it's certainly going to be a lot of fun in the next few years to work that out."

9

Where Did the Universe Come From? Why Hasn't It Curled Up into a Ball?

THERE ARE two traditional answers to the first question. God may have created the universe. Or perhaps there was no creation; the universe may always have existed. Both answers have respectable histories. For example, Plato recounts a creation myth in the *Timaeus* while Aristotle suggests that the second possibility is the more reasonable.

On the other hand, physicists consider both answers to be unsatisfactory. This has nothing to do with their religious beliefs, or the lack of them. Scientists seem to be no more religious or irreligious than other people; many are practicing Jews or Christians, many adhere to other religions, and still others are agnostics or atheists.

The reason that the first answer is deemed to be unsatisfactory is that scientists do not like to invoke the idea of God to explain natural phenomena. Saying that the universe exists because God created it would be like saying that gold is yellow because God ordained it to have that color. Naturally it is perfectly possible that God (if there is a God) did precisely this, but to say so is not to engage in scientific explanation. Whatever God did or didn't do, gold must be yellow because there exist physical laws that cause it to reflect light at certain wavelengths and not others.

Plato to the contrary, the idea that there had been no creation, that the universe had existed eternally, was quite common among the Greeks of the classical era, but scientists would feel even more uncomfortable giving this kind of answer. At the very least, they would feel obligated to make the concept seem reasonable by engaging in theory or scientific speculation. There does seem to have been a big bang, after all. Consequently any scientist who wants to maintain that the universe had always existed in one form or another has to explain what was going on before the big bang took place.

You might think that speculation about such a subject as the origin of the universe would lie in the province of metaphysics or theology, not science, and indeed, at one time this was the case. It isn't so long since scientists tended to be skeptical of the idea that we could even say anything very meaningful about the conditions that prevailed during the first few seconds after the big bang. To suggest that we could go back even farther and say anything about the origin of the universe would have been thought absurd.

During the 1970s and 1980s, however, attitudes changed as physicists discovered that events that took place within the first tiny fraction of a second after the beginning of the big bang could have observable effects today. In particular, many scientists became convinced that an inflationary expansion that began at a time of 10^{-35} seconds produced features in the universe that could be observed today.

As I have noted previously, there is nothing that would

compel us to believe that an inflationary expansion took place. However, the inflationary paradigm has been so successful that it is hard to imagine that it could be completely inaccurate. In fact, we should not find it surprising that scientists should be so encouraged by the success of the inflationary universe theories that they would attempt to look back to even earlier times and to speculate about the origin of the universe itself. In a sense, some physicists have even begun to speculate about what might have been going on before time began.

From time to time in this book, I have talked about the frontiers and boundaries of science, using the term *frontier* to indicate research that proceeds from reasonably well-established theory. Theoretical research based on the well-confirmed big bang theory can be said to lie on the frontier, for example. The inflationary universe paradigm, which is plausible, but less well confirmed, can be said to lie near the boundaries of science.

If I continue to use this terminology, then it is probably necessary to say that speculation about the origin of the universe sometimes lies outside of science altogether, or at least beyond the boundaries. For when scientists speculate about matters such as this, they are beginning to explore areas of thought where they have no theory to guide them.

When scientists use Einstein's general theory of relativity to trace back the expansion of the universe to the beginning of the big bang, they cannot avoid coming to the conclusion that, at the beginning, all the matter in the universe must have been compressed into a mathematical point called a singularity. In other words, according to general relativity, at time zero the universe had zero spatial dimensions, and the density of matter was infinite. Furthermore, there is no way that such a conclusion can be avoided. During the 1960s, the British physicists Stephen Hawking and Roger Penrose proved a series of mathematical theorems that indicated that, if general relativity is a wholly correct theory, then the conclusion that there was an initial singularity is unavoidable.

Was the density of matter at the beginning really infinite? Of course not. All scientific theories have their limits. There are always extreme conditions under which the best of them break down, and when infinities begin to crop up in a theory, they can generally be taken as signs that those limits have been reached. The prediction of a singularity, in other words, is most likely an indication that we have ventured into an area where general relativity is no longer valid.

There is nothing very surprising about such a conclusion. After all, during the early stages of the big bang, matter must have been in a very highly compressed state. Consequently, quantum effects must have been important. We would have to have a theory of quantum gravity to have any hope of correctly describing what was going on, but, as we have seen, a theory of quantum gravity does not yet exist. Many physicists hope that superstring theory will eventually become one. However, superstring theory is still in an early stage of development. It has not yet reached the stage where it even comes close to predicting the behavior of an individual proton, neutron, or electron. It may be decades before it is even possible to imagine applying it to the early stages of the big bang.

We cannot use the standard model to describe the early stages of the big bang either. The standard model works well enough when the force of gravity can be neglected, which is the case under ordinary terrestrial conditions, and under conditions that exist in stellar interiors or in high-energy particle accelerators, where gravity is so weak that we never have to take it into consideration. However, during the initial stages of the big bang, particles were so close together that their gravitational interactions must have become important indeed. In fact, it is possible to compute that, during the first 10^{-43} seconds, gravitational forces must have been as strong as those produced by the other three interactions.

We must conclude that, when we attempt to look back to the origin of the universe, all the known laws of physics break down. During the first 10^{-43} seconds, general relativity no longer works, and the standard model can't be applied either.

However, somewhat paradoxically, this has not proved to be an impediment to speculation. In recent years, physicists have attempted to bypass this theoretical barrier. They have attempted to delve into such matters as where the universe might have come from, and what might have been happening before the beginning of time.

THREE THEORIES OF THE ORIGIN OF THE UNIVERSE

There exist numerous theories about the origin of the universe, or at least some of these theories have numerous different variations. However, the theories can be classified into three different types:

1. Although the universe is not eternal, it has neither a beginning nor an end. Or, as Stephen Hawking has put it, the universe might be finite and yet have no boundary. It might have begun in "imaginary time."
2. The universe began as a microscopic quantum fluctuation. It popped into existence out of nothing, just as virtual particles do.
3. Our universe began as a quantum fluctuation in a previously existing universe. Obviously, this is a variation of 2. However, I have listed it separately because it implies that universes may reproduce themselves endlessly.

NO BOUNDARY

The idea of a universe that is finite but has no boundary was developed by Stephen Hawking in collaboration with physicist James B. Hartle of the University of California at Santa Barbara. This proposal—Hawking is at pains to emphasize that it is only a proposal, not a full-fledged theory—is described in Hawking's book, *A Brief History of Time*.

To tell the truth, I find Hawking's account of the theory in this book somewhat confusing. It wasn't until I read a somewhat more technical account of it elsewhere that I felt I understood what Hawking and Hartle are really saying. Furthermore, I found Hawking's use of the concept "imaginary time" in his book to be potentially misleading to the layperson, because he does not distinguish between the meanings of the word "imaginary" in ordinary language, and the very different mathematical concept of an imaginary number.

My explanation of the Hawking-Hartle hypothesis will therefore seem somewhat different from the one in Hawking's book. It may be that the reader will decide that it is my account that is confused, while Hawking's is more than sufficiently clear. I should point out, however, that my account is based on descriptions of the proposal that Hawking has published elsewhere, and that it can therefore be taken to be reasonably authoritative.

Hawking points out that, if we imagine that the universe was created at some particular point in time, certain unanswered questions remain. If the universe was not initially an infinite-density singularity, then it must have had some particular initial state. However, the laws of physics cannot tell us why it should have been in one particular state and not another. The known laws of nature only tell us how the universe evolved afterward.

Like many physicists, Hawking and Hartle would prefer to believe that the laws of physics are ultimately capable of explaining anything that we can observe. Consequently, they attempted to see if it was possible to imagine that the universe had no beginning, and thus avoid the initial-state problem.

Now, one obvious way of doing this would be to follow the ancient suggestion of Aristotle and to postulate that time extends backward into the infinite past. However, this doesn't really solve anything. We are still confronted with the impossibility of saying why the universe should have had some particular properties at some particular time. Even if something has been going on forever, we would still want to ask what properties it had in the past to make it what it is today.

So Hawking and Hartle set aside the initial-condition problem, and asked instead what effects quantum mechanics might have had on the nature of space and time. They found that, when the universe was very young, and space was very compressed, the quantum uncertainties associated with the Heisenberg uncertainty principle would have begun to erase the distinctions between space and time. If one went back far enough, time might become "spatialized." The universe would no longer have three dimensions of space and one of time. On the contrary, it would be something that existed in a kind of four-dimensional space.

A four-dimensional space can curve back upon itself to form a closed surface that has no edges or boundaries. This would be the analogue of a two-dimensional surface that closes upon itself to form a sphere; it is also analogous to a closed universe. But there is an important distinction. The Einsteinian closed universe possesses only three spatial dimensions. In a closed universe, the three spatial dimensions close upon one another, while time remains something resembling a straight line (Einstein once compared this universe to a cylinder). In the universe of Hawking and Hartle, on the other hand, four dimensions, not three, close upon themselves.

The proposal can therefore be summarized as follows: When we go far back enough in time, there is no longer any time, only four spacelike dimensions. Consequently the universe has no beginning, for the simple reason that time no longer has a timelike character. This theory does not conflict with any known facts, and it is perfectly consistent with the inflationary paradigm. Presumably, by the time the inflationary expansion began, time had become time as we know it.

If the Hawking-Hartle universe has no beginning, neither does it have an end. There is no boundary to time in the future either because exactly the same phenomenon takes place.

If this proposal is correct, the universe would have to be closed. It would have to have a mass density high enough that the expansion of the universe would eventually halt, and a contracting phase begin. Eventually the universe would become so compressed that quantum effects would again become important. The time dimension would then become spacelike, and the universe would again have four spatial dimensions with no edge or boundary.

What happens after that? There is no such thing as "after that." This is an expression that refers to the passage of time, but there will be no time, at least not in the sense that we understand the term. Asking what happened before the big bang, or what will happen after the final contraction would be, as Hawking puts it, "like asking for a point one mile north of the North Pole."

Something Out of Nothing

Since, as Hawking himself admits, the idea that space and time might be unbounded and finite is only a proposal, one that cannot be deduced from any other principle, we are free to ask if there might not be other plausible ways in which the universe could have begun. As it happens, there are several.

One of the simplest is the idea that the universe might have begun as a quantum fluctuation in which some virtual particles were created out of empty space. This creation scenario, which was proposed by physicist Edward Tryon in 1973, was discussed in chapter 2 in connection with the inflationary universe paradigm.

However, Tryon's hypothesis is really independent of the

inflationary universe theories. The basic idea is that, if the total mass–energy content of the universe is zero (recall that the total gravitational energy is negative), then Heisenberg's uncertainty principle tells us that it could exist for an indefinite period of time. This is true whether there was an inflationary expansion or not.

There are a number of variations on Tryon's idea. For example, in 1978, four Belgian physicists, R. Brout, P. Englert, E. Gunzig, and P. Spindel, suggested that the universe might have begun with the creation of a particle-antiparticle pair with masses of 10^{19} GeV each. Once this pair of supermassive particles existed, the production of other particles of matter would have been stimulated. Presumably the process continued until the inflationary expansion began, filling the rapidly expanding universe with still more matter and energy.

Then, in 1981, physicists Heinz Pagels and David Atkatz of Rockefeller University suggested that the universe might have begun, not with the creation of a pair of particles, but with a sudden change in the dimensionality of space. According to their theory, space—which contained no matter originally—initially had a large number of dimensions. The universe, Pagels and Atkatz said, might have begun with a change in the quantum energy state of this space. They hypothesized that spacetime could have suddenly "crystallized" into the ten dimensions of superstring theory.

In 1983, Alex Vilenken of Tufts University went one step further, and suggested that the primordial chaos from which the universe was created didn't even have a definite dimensionality. In Vilenken's theory, the very concept of the dimensionality of spacetime only took on meaning after the universe came into existence.

In a way, all these theories are reminiscent of a creation myth found in a number of different cultures in the ancient Middle East. According to this myth, the world was created, not out of nothing, but out of a kind of formless chaos. We find echoes of this myth in the second verse of the first chapter of Genesis, where we read, "And the earth was without form, and void; and darkness was upon the face of the deep." Naturally I am not suggesting that the author of Genesis or the creators of ancient mythology had any premonitions concerning modern physics. However, it is interesting that the idea of creation out of a primordial chaos should suddenly be reborn in a new form today.

SELF-REPRODUCING UNIVERSES

The theories that I have outlined above are very speculative. The authors of these theories do not make any attempt to show that the universe *did* pop into existence out of nothing, nor do they demonstrate that the universe *could have* come into existence this way. No one really knows whether the laws of physics would allow universes to be created in this manner or not. All that has really been demonstrated is that there is nothing intrinsically implausible about the idea. In other words, we know so little about the origin of the universe that no one can say that it was *not* created in this manner.

Once speculation has been taken this far, there are no impediments to progressing a little further, and asking how many universes there are. Is there only one? Or are there many universes, perhaps even an infinite number? There doesn't seem to be anything intrinsically implausible about this. If a universe can be created once, such an event could presumably happen many times.

It sounds a little illogical to speak of many universes, since the word "universe" is generally used to refer to everything that exists. Therefore, before I go on, it might be best to introduce some new terminology so that no semantic difficulties arise. From now on, I will use the word *universe* to mean a self-contained region of spacetime, such as the universe in which we live. If I need a word to describe the entire ensemble of universes that may constitute all of reality, I will use the term *cosmos* instead.

If universe creation is an event that happens over and over again, it could take place in two different ways. A universe could be created in a spacetime that had no connection with the spacetime of existing universes. Alternatively, new universes could be created out of the empty space within universes that already exist. In other words, it is possible that universes might reproduce themselves.

If universes are created in separate spacetimes, and never have any connection with one another, we will never be able to tell whether or not any universes other than our own exist. We could not even say "where" they were if they did. "Where," after all, is a word that refers to position in spacetime, and the spacetimes of these other universes would have no connection with that of our own. We could even ask if it was philosophically meaningful to speak of the "existence" of such universes. If something cannot, in principle, ever be observed, can one really say that it "exists"?

The idea that universes might reproduce themselves, that they might begin as quantum fluctuations in previously existing universes, seems to be a much more fruitful hypothesis, since it could easily have observable consequences. There might be some way that we could see universes being born.

Reproducing universes would presumably have to be closed. At least it would be difficult to conceive of the creation of an infinite, open universe within a previously existing one. But if universes can reproduce, wouldn't we expect to see universes being created within our own? And wouldn't a rapidly expanding universe created within ours eventually engulf us? The answers to these questions seem to be no. Einstein's general theory of relativity implies that such a universe, when seen from inside, might seem to be rapidly expanding, and yet appear to be an object very much like a black hole when seen from outside.

Indeed, we might note that there is a sense in which a closed universe very much resembles a black hole, which is so massive and compressed, and has gravity so strong, that nothing, not even light, can escape from it. Nothing can escape from a closed universe either, and if our universe is closed, then we could think of it as a black hole within some larger universe, which could in turn be embedded in yet another universe.

However, if universes do reproduce themselves, a newly created universe would not necessarily remain within its parent universe. For example, one that was created in our universe might "pinch off" from our spacetime and disappear. For a brief moment of time, a thin strand of spacetime called a wormhole might connect the two universes, and then quickly vanish.

The idea of self-reproducing universes has been elaborated by Soviet physicist Andrei Linde in his chaotic inflationary universe theory. Linde builds upon previous inflationary theories, and speculates that inflationary expansions are continually coming into being in numerous self-reproducing universes. In some of these universes, the inflationary regime never ends; they continue expanding at this fantastically rapid rate forever. In other universes, like our own, the rate of expansion diminishes to a more leisurely pace like the one that astronomers observe today. New inflationary universes are being created all the time within the ones that exist already. Soon after they are created, they bud off and separate, and then give birth to universes of their own.

In Linde's scenario, some universes might eventually enter into a stage of contraction, and eventually crush themselves out of existence in a big crunch. However, since any one universe could give birth to numerous others, the cosmos would go on forever. Our universe might not be eternal, but the cosmos would be.

Linde's theory has one other very interesting implication. There is really no reason why the laws of physics, or even

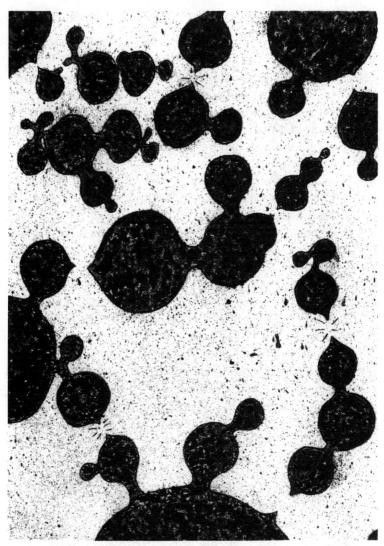

Self-reproducing universes. The cosmos may contain a very large, possibly infinite, number of separate, self-contained universes. Furthermore, it is possible that these universes reproduce themselves by a kind of "budding" process. Here, an ensemble of universes is reproducing in just such a manner. After the daughter universes form, and begin to undergo inflationary expansions, they may break off and sever all connections with their parent universes. Naturally, this hypothesis of self-reproducing universes is very speculative. There is no evidence to indicate that such a process actually takes place in nature.

the dimensionality of space, would have to be the same in all universes. It is possible that there could be a kind of "genetic code" that would cause daughter universes to resemble their parents. Even in such a case, there could presumably be "mutations." Our universe, for example, could be a mutated descendant of a universe in which the laws of physics were just different enough that the creation of life was impossible.

You might think that speculation of this sort is so "far out" that it could have no possible relation to reality. However, this might not be the case. In fact, the existence of other universes could have significant implications. For example, suppose that physicists established that superstring theory was correct, and that superstring theory did not uniquely determine all the observed laws of physics. In such a case, it could still be a "theory of everything," but one that allowed a number of different possibilities. Furthermore, it could turn out that these various different possibilities were all realized in one universe or another.

Finally, there is one other possibility that might be mentioned, one that has something of a science-fiction character, but which is not entirely outside the realm of possibility. If universes do begin as tiny quantum fluctuations, it is conceivable that the creation of new universes might be within the capabilities of an advanced technology. This raises the question of whether our universe could have been created in this way. If it could have been created deliberately, can we even be sure that the universe we see is not the result of some graduate student's experiment gone awry? Naturally, when I say this, I am making a joke. However, it is somewhat sobering to realize that this is not completely impossible.

Einstein's Blunder

When Albert Einstein published his general theory of relativity in 1916, he gave the world a set of equations that described the workings of gravity and the curvature of space. Initially, he had been concerned with working out a theory that showed how gravity worked. Obviously, such matters as the implications of his theory for an understanding of the structure of the universe could be worked out later.

After the theory was published, Einstein set to work seeking solutions to his equations that would describe the entire universe, but soon realized that his theory seemed to imply that the universe had to be either expanding or contracting. When he obtained this result, he was disturbed, for in 1916 no one had ever heard of an expanding universe. It had always been assumed that the universe was static. For that matter, at this time, most astronomers still believed that our Milky Way galaxy *was* the universe.

Believing that he had to find a solution that would correspond to a static universe, Einstein looked for a way to fix up his theory. It didn't take long to find one. He observed that his theory allowed the introduction of a quantity, which he called the cosmological constant, that could be positive, negative, or zero. If he assumed that this constant was just the right size, Einstein observed, a static universe would result.

Einstein's assumption concerning the cosmological constant was tantamount to assuming that a kind of repulsive, antigravity force existed in the universe, and that this force could balance out gravitational attraction at large distances. No such force had ever been observed, but that was really no argument against its existence. After all, the theory implied that space was curved, and no one had yet observed curved space either.

In 1917, Einstein published a paper in which he described his conception of the universe. The universe was closed and finite, he suggested. Furthermore, the cosmological antigravity force could ensure that it always retained the same dimensions.

Einstein's paper proved to be wrong on two counts. In the first place, the assumption that such a universe would be static proved to be a blunder. Other scientists soon showed that Einstein's universe was unstable. The cosmological constant and gravity could only balance one another if the dimensions of the universe were precisely right. If it expanded by just a tiny amount, gravity would weaken slightly, the repulsive force would become dominant, and the universe would grow bigger and bigger. On the other hand, a very minute contraction of the universe would give gravity the upper hand, and the contraction would continue unabated, and would in fact accelerate as matter became more compressed and gravity grew progressively stronger. Einstein's universe, in other words, was like a pencil balanced on its point: it could not remain static because it would soon topple one way or the other. Twelve years after Einstein's paper was published, it was demonstrated that the universe was not static after all. In 1929, Hubble announced his discovery that the universe was expanding.

Einstein was later to refer to the introduction of the cosmological constant as "the greatest blunder of my life." Nowadays cosmologists are not so sure that it was a mistake at all. Many of them think that the constant should be left in Einstein's equations. They feel, furthermore, that if the constant turns out to be zero, it is this fact that must be explained; a value that was exactly zero would be surprising indeed.

The introduction of the cosmological constant has parallels in many of the theories of physics. When scientists work out theories mathematically, they often find that certain numbers, called constants of integration, can be added in a natural way. In fact, in most cases, it would be a mistake to leave these constants out: with them, a theory becomes more general; without them, it can usually only be applied to special cases.

Measuring the Constant

The cosmological constant cannot be left out of the equations of general relativity on *a priori* grounds; standard mathematical practice requires that it be retained. Furthermore, it should not be set equal to zero unless it is measured to be zero. Physics is an empirical science, and such quantities should be determined experimentally.

This does not mean that scientists should set up apparatus to measure tiny attractive or repulsive forces (they can be either, depending upon whether the constant is negative or positive) that emanate from distant regions of the universe. The gravitational attractions of distant objects cannot be measured directly, nor could cosmological force be observed either.

However, if a nonzero cosmological constant exists, it should affect the motions of distant galaxies. A positive constant, corresponding to a repulsive force, would tend to make galaxies move away from one another more rapidly, while a negative constant would correspond to an attractive force that would slow down the expansion of the universe. Furthermore, these effects would be distinguishable from the gravitational forces produced by matter in the universe. This would be true even in the case that a negative constant produced an attractive force. Gravity becomes weaker when a gravitating body is farther away. The cosmological force does not depend upon distance. Thus the two should produce different kinds of "braking" effects upon the expansion of the universe. Furthermore, it should be possible to distinguish one from the other.

As I have pointed out several times by now, when astronomers study galaxies billions of light-years away, they are also looking back billions of years into the past. Thus they can compare the rate of expansion of the universe in past eras to that observed today, and determine whether gravity alone can account for the changes that have taken place.

Galactic surveys have been conducted out to distances of 10 billion light-years, and no evidence for the existence of a cosmological constant has ever been found. If one does exist, it is so small that it has had no discernable effects over the last 10 billion years.

A Discrepancy of 10^{120}

You might think that astronomers and physicists would conclude that the matter had been put to rest, that we could conclude that the cosmological constant was zero, and drop it from the equations of general relativity. Unfortunately, this cannot so easily be done, for there exist theoretical reasons for believing that the cosmological constant should be large indeed. As a matter of fact, it should be so large, and the resulting force so strong, that the entire universe should long ago have curled up into a tiny ball with a diameter less than that of an atom.

As we have seen, if such quantum field theories as QED and QCD are to be believed, "empty" space is never really empty, but is on the contrary full of seething activity. The "emptiness" of the vacuum is filled with quantum fields, and with enormous quantities of virtual particles that are constantly being created and destroyed.

Furthermore, there is an energy, which can be calculated, associated with all this activity. When the calculation is performed, the self-energy of the vacuum turns out to be enormous. Since energy and mass are equivalent, this energy should have significant gravitational effects. In fact, this vacuum energy should create a force precisely like that associated with the cosmological constant. Although it would be gravitational in nature, it would not vary with distance as, for example, the gravitational attraction of a galaxy would, since "empty" space is everywhere.

Furthermore—and this is the problem that cosmologists find so baffling—this force should be about 10^{120} times larger than the maximum cosmological force that is consistent with observations. The vacuum energy should create a cosmological constant so large that the universe should never have been able to expand beyond microscopic dimensions.

There are a number of different contributions to the theoretical energy density of the vacuum. The virtual particles predicted by the theories that make up the standard model make a contribution to it, and the fields associated with the hypothetical Higgs particles make an even larger one. If there exist as-yet-undiscovered elementary particles, they would make a contribution too. The larger the number of different kinds of particles, the more numerous the virtual particles that can be created, and every additional variety of virtual particle increases the vacuum energy still more.

Since it is not known what particles will be discovered in the future, or how the standard model might be modified, the vacuum energy cannot be calculated precisely, though estimates of its minimum value can be made. These estimates indicate that it should theoretically be at least 10^{120} times larger than the maximum value that is consistent with observations. Even if we ignore all the quantum particles and fields except those associated with the strong force and with quarks, the theoretical value obtained is still far too large. If quarks and gluons were the only particles that existed, the vacuum energy density would still be too large by a factor of 10^{41}. Though that sounds better than 10^{120}, it can hardly be said to bring about an agreement between theory and observation—10^{41} is a hundred thousand billion billion billion.

Why There Is Nothing Rather than Something

Naturally, attempts have been made to rectify this discrepancy. In fact, in 1988, Harvard University physicist Sidney Coleman published a paper entitled "Why There Is Nothing Rather than Something," which caused quite a stir within the theoretical physics community. In this paper, Coleman presented a hypothesis that seemed capable of explaining why the observed cosmological constant should be zero. Though Coleman's idea was only a conjecture, it at least provided a plausible explanation where none had existed before.

Before I explain Coleman's conjecture, it is necessary to digress a little, and to comment on some of the recent speculations of Stephen Hawking. Hawking's book *A Brief History of Time* covered his work only up to around 1985, the point at which the first draft was finished. As a result, there is no mention in it of some of his more recent theoretical work.

Some of this work is related to the concept of reproducing universes discussed earlier in this chapter. In particular, Hawking has carried out theoretical investigations of the possible effects of the wormholes that might connect the newly created baby universes* to our own.

In principle, these wormholes could be of any size. However, ones that were very much larger than about 10^{-33} centimeters in diameter would be relatively improbable. Now 10^{-33} centimeters is about 10^{20} times smaller than the diameter of a proton. We could no more observe such a wormhole than we could see a superstring (which is supposed to be about the same size), and if such a wormhole could be observed, we would not see it for long. It would pop into existence and then disappear again in a time of about 10^{-43} seconds.

It doesn't necessarily follow that the existence of such wormholes would have no observable effects. According to Hawking, such effects could be dramatic indeed. Hawking begins by noting that every now and then a particle from our universe, such as an electron, would disappear into such a wormhole, while an identical particle from another universe would come out of it. The requirement that the identical particle emerge from the wormhole is a consequence of certain basic laws of physics which require the conservation of such quantities as mass and electric charge. Unless there is some compelling evidence to the contrary, we should as-

* "Baby universe" sounds like a "cute" colloquialism. Nevertheless, it really isn't; in fact, it is rapidly becoming part of accepted scientific terminology.

sume that these laws will be observed whether our universe is connected to other universes or not. In any case, no one has ever observed a single electron suddenly to appear or disappear.

Hawking imagines that our universe is filled with astronomical numbers of wormholes that are constantly flickering in and out of existence. Thus the particles that make up our world are constantly falling into wormholes that we never see, while particles from other universes take their place. Naturally, we are never aware that any of this is happening. As far as we are concerned, particles continue to behave as though these other universes didn't exist. An electron's trajectory* cannot be altered when it changes places with its alternate-universe partner, for this would violate other well-established laws of physics, the laws of conservation of energy and momentum.

If such particle exchanges had no observable effects whatsoever, it would be senseless to speak of them, for physics is not a science that deals with phenomena we cannot see, or with effects we cannot measure. But Hawking does not say that these exchanges would have no effects. On the contrary, he suggests that they might affect our measurements of the electron mass, and the masses of other particles as well. He points out that if particles can disappear into, and emerge from, wormholes, then they will seem to have masses that are greater than those of particles which always remain in the same universe. Furthermore, wormhole exchange could have similar effects on a particle's observed charge.

Once he has established this result, Hawking goes on to suggest that wormholes might be responsible for *all* particle masses. He suggests, furthermore, that these wormholes might play a role in all processes in which one particle seems

---

* Strictly speaking, the indefiniteness associated with Heisenberg's uncertainty principle prevents us from speaking of a "trajectory." So it would be more precise simply to say that the electron's apparent behavior is not altered.

to interact with another. For example, if an electron and a virtual photon disappear into a wormhole together, the effect will be as though the (force-carrying) photon had been absorbed by the electron.

At this point, however, Hawking's theory seems to run into trouble, for detailed calculations seem to predict that wormholes would produce particle masses about 10^{20} times larger than the mass of the proton. It seems that Hawking's fantastic, but attractive, hypothesis has given an absurd result. At least, physicists might have drawn this conclusion in former times. However, nowadays, when superstring theo-

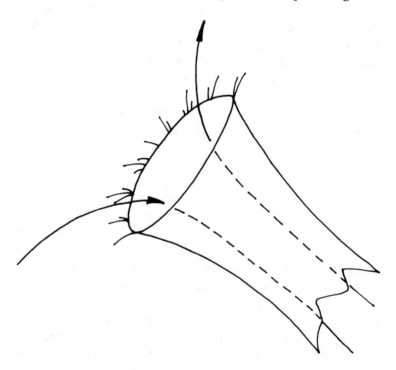

The origin of mass? According to Hawking's hypothesis, subatomic particles such as electrons may be constantly disappearing into wormholes and traveling to other universes. In this diagram, an electron enters a wormhole while an identical particle leaves it and enters our universe. According to Hawking, electrons could acquire mass by this process.

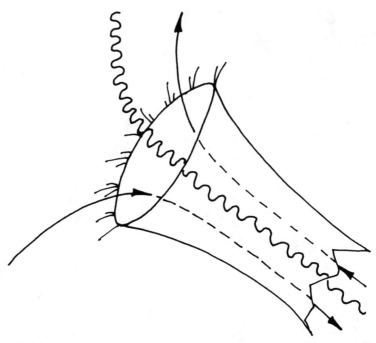

A process similar to that depicted in the figure on page 188 above may be responsible for particles' charge. In this diagram, an electron enters a wormhole, while an electron and a photo emerge.

rists find that their theory predicts particles with masses 10^{19} times larger than protons', and when cosmologists deal with a cosmological constant that appears to be 10^{120} times smaller than it should be, 10^{20} doesn't seem so terribly fearsome at all. Of course, if the theory is ever to be taken seriously, the discrepancy will have to be eliminated, but this is precisely what a number of theoretical physicists at a number of universities are attempting to do. They are seeking to discover ways in which the theory can be modified to give more reasonable results.

If particles do acquire such properties as charge and mass in this manner, it doesn't necessarily follow that these properties must be the same in different universes. These "constants" of nature could vary in a random way from one

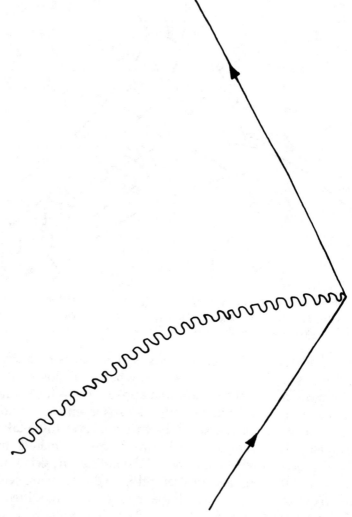

The wormholes that Hawking postulates could not be seen because of their submicroscopic dimensions. The process shown in the figure above would therefore
appear to be one in which the electron remained in our universe and emitted a
photon as it traveled through space.

universe to another. Though the mass of the electron was a consequence of its passing into and out of wormholes, it could conceivably take on a different mass when it appeared somewhere else.

I must emphasize once more that all this is very speculative. No one is really sure that wormholes or alternate universes exist, or that particles could travel from one universe to another if they did. If any of this did turn out to be correct, the idea that certain quantities vary randomly from universe to universe could turn out to be wrong.

However, as Coleman points out, if all these ideas turn out to be correct, a great deal would be explained. In particular, if the constants of nature vary in a random way, there is no reason why the cosmological constant could not exhibit such random variation also. If this were the case, the wormhole effect would cause some universes to have very large cosmological constants, while others possessed very small or zero constants.

Furthermore, according to calculations that Coleman has carried out, universes with a cosmological constant of zero would be much more probable than those in which it had any other value. It is possible that our universe has not curled up into a tiny ball simply because this fate is experienced by very few universes. Similarly, it is possible that the vacuum energy does not create dramatic effects simply because spacetime is full of holes.

DARK MATTER

It cannot be emphasized too strongly that years, and perhaps decades, of work will be required before any of the ideas discussed above can be regarded as anything but wild speculation. It is also possible that, in the end, they will be discarded (and perhaps be replaced by even wilder concepts). Perhaps it would not be a bad idea to close this chapter by returning to a topic that is, by comparison, rather mundane: the problem of the nature of dark matter.

The dark matter problem can be summed up as follows:

1. Astronomical observations have established that the universe contains mass that we cannot see. No one is quite sure exactly how much of it there is.
2. The plausibility of the inflationary universe paradigm makes it seem reasonable to believe that the density of matter in the universe is very, very close to the critical value.
3. The luminous matter that exists in stars and galaxies provides a matter density only about 1 percent of the critical value. If we add other kinds of baryonic matter that might exist—dim stars, Jupiter-size objects, and so on—then baryonic matter can't provide much more than about 10 percent of the critical density.
4. Consequently, it is generally assumed that 90 percent of the mass of the universe exists in the form of exotic particles left over from the big bang, or in the form of mass-carrying neutrinos, or as cosmic strings.

Now suppose that the assumption in paragraph 4 turns out to be wrong. Suppose it is established either that these objects do not exist, or that there is not as much mass contained in them as scientists think. In such a case, what could the dark matter be?

In an article in the March 1989 issue of *Physics Today*, physicist Bertram Schwarzschild suggested a possible answer to this question. Perhaps, Schwarzschild said, the nonbaryonic dark matter didn't exist at all. Perhaps there was a cosmological constant too small to be measured, but large enough to mimic the effects of dark matter. The vacuum energy density that produced the cosmological constant would have a mass equivalent after all. Perhaps the dark "matter" was really energy instead.

A universe with a nonzero cosmological constant,

Schwarzschild pointed out, would differ from a universe with a zero constant in one important respect. The amount of time that had elapsed since the big bang would be greater; the expansion of the universe would have been "braked" by the cosmological force in a different manner than by an equivalent amount of mass. The universe could then be assumed to be much more than 10 or 15 billion years old, and astronomers would no longer be confronted with the problem of having to explain why the oldest stars they observed appeared to be as old as, or older than, the universe itself.

IV

THE
FRINGES
AND THE
EDGES OF
SCIENCE

10

On the Edge

SUPERSTRING THEORY and theories about the origin of the universe represent the farthest boundaries of science. However, a discussion of these topics does not really exhaust my subject, for I have not yet considered the fringes and edges of science.

When I speak of "fringes," I am thinking of pseudoscientific theories, crackpot ideas, and scientific hypotheses based more on wishful thinking than empirical observation. Although this topic has a certain amount of interest, I do not propose to delve into it in this book. Here, I am concerned with examining only those ideas that can, in some sense, be called truly scientific. The ideas that we find on the fringes of science generally do not have this character. In any event, I have written about fringe ideas elsewhere; some are discussed at length in my book *Dismantling the Universe* (see Bibliography).

However, I do want to pursue the idea that there are certain kinds of thought that lie on the "edges" of science. I think it is possible to classify such thought in two broad subcategories. On one hand, there are kinds of speculation, engaged in by scientists, that are more philosophical than scientific in nature. On the other hand, there is speculation that has no solid theoretical or experimental justification.

Thought in the latter category can be distinguished from pseudoscience because there are often good scientific motivations for engaging in this kind of speculation. In fact, some of the theories about the origin of the universe fall into this category, or at least they exist in an area where the "boundaries" and "edges" of science blur into one another.

If we are to discover what *is* true, it is necessary to determine what *might* be true. Science would not advance if scientists voluntarily limited their intellectual horizons. Consequently, physicists sometimes propound ideas for no other reason than to show that these ideas are not inconsistent with the known laws of physics.

TACHYONS

A good example of this kind of speculation would be the hypothesis proposed by the American physicist Gerald Feinberg that there might exist particles that could travel at velocities greater than the speed of light.

According to Einstein's special theory of relativity, no particle with nonzero mass can be accelerated to light velocity. The theory says that the mass of any particle or material object must increase as it is accelerated to high velocities. The heavier the particle becomes, the harder it will be to accelerate it further; it will have more inertia. A greater quantity of energy is required to achieve each successive increment in velocity, and an infinite amount of energy would be required to attain the speed of light.

Naturally, Feinberg was very much aware of this aspect of Einstein's theory. However, he was not suggesting that anything could be accelerated past the light "barrier." He simply noted that the existence of faster-than-light particles, which he called tachyons, would not contradict Einstein's theory if he assumed that they encountered the same barrier from the other side. Tachyons, in other words, could conceivably exist if they always maintained velocities that were greater than that of light.

At first, the idea of tachyons elicited a certain amount of theoretical interest, and attempts were made to detect them experimentally (a tachyon could have been identified by the fact that it traveled backward in time), but eventually the interest faded. Nowadays, if a theory predicts the existence of tachyons (some superstring theories do this, for example), this is considered to be a serious defect.

The idea that there might be faster-than-light particles was initially viewed with interest. Then it was tested, and finally discarded; but it was not discarded because any flaws were found in Feinberg's reasoning, or because experiment proved that tachyons did not exist. In fact, proving the non-existence of something experimentally is impossible. Tachyons ceased to interest scientists because the hypothesis of their existence seemed to have no significant experimental or theoretical consequences. There were no observed phenomena that could be attributed to the existence of tachyons, and there were no promising theoretical ideas that required the existence of tachyons if they were to seem plausible.

In other words, the idea of tachyons simply came to seem irrelevant—or perhaps worse than irrelevant, for the hypothesis not only didn't solve any outstanding theoretical problems, it created new ones. If tachyons had turned out to be real, scientists would have had to deal with particles moving backward in time, and would have had to explain how the past could be influenced by the future. To be sure, relativity says that faster-than-light particles will only seem to some observers to be moving backward through time, but then allowing even some observers to catch a glimpse of the future would be bad enough.

Yet it should not be concluded that Feinberg's hypothesis was a silly one—it was anything but that. If the idea hadn't been explored, no one would have known whether it was reasonable or not when it was encountered in superstring theories, and the meaning of the light "barrier" in relativity would not have been as well understood. Furthermore, there have been other, equally bizarre, ideas that have turned out either to be true, or to be very promising. If scientists stopped considering concepts that seemed to be strange, then research in particle physics and cosmology would have slowed to a crawl long ago (and I would not be writing this book).

BIZARRE IDEAS

When physicists map out the edges of science, they are engaging in an activity somewhat different from the work of those in the scientific mainstream. Exploring the edges of science is an attempt to discover what physical reality might possibly be like, not an attempt to work out the details of what it is. We could say that mainstream and frontier science seek to map out the known world, while edge science tries to understand what kinds of worlds might be possible.

The scientists who work on the edge are often concerned, as Feinberg was, with trying to find out what kinds of phenomena might just be consistent with known physics. Some of them have asked, for example, whether time might not run backward in a contracting universe (why not, if it runs forward in an expanding one?), whether there might not be an infinite number of alternate universes, and whether a positron might not be an electron that moves backward in time. For that matter, if a positron is a backward-moving electron, for all we know, there might be only one electron in the universe. What we perceive as many particles could be the same one going back and forth past us in both directions.

Such conceptions often seem crazy, but they are no more bizarre than other ideas that have come to be accepted. A list of such "crazy" conceptions might include the existence of virtual particles, the Einsteinian idea that gravity can bend a ray of light, and the idea that there exist such objects as black holes. Although these ideas are now familiar, they all seemed difficult to believe when first suggested. For that matter, even the ideas that atoms had constituents or that the universe was expanding seemed bizarre at one time.

Scientific progress often depends upon a willingness to give up commonsense ideas. Good physicists must be willing to discard ingrained outlooks and prejudices; they must ask, not what the limits of the universe seem to be, but what they could be. They must ask what kinds of bizarre phenomena the laws of physics might allow, and think about the seemingly impossible in order to find out what reality is.

BLACK HOLES, WORMHOLES, AND TIME TRAVEL

With these ideas in mind, I will describe a theory that is in all likelihood not true, but which just conceivably might be. Strictly speaking, we should probably not call it a "theory" at all, since it is not an investigation of the laws of physics so much as speculation concerning the things that a highly advanced technological civilization might be capable of doing.

In particular, the physicists who have developed this "theory" ask whether the beings who had developed such a civilization might not be capable of two things often described by science-fiction writers: to travel across interstellar space at velocities greater than that of light, and to engage in time travel.

Before I describe the theory, which was developed by California Institute of Technology physicists Michael S. Morris, Kip S. Thorne, and Ulvi Yurtsever, I must give a little background, and discuss some ideas that—like Feinberg's tachyon hypothesis—aroused a great deal of interest at one time, but which were later discarded as unrealistic.

According to the general theory of relativity, a black hole has two important features, called the event horizon and the singularity. The event horizon is a spherical surface through which objects can travel in only one direction. There is nothing to prevent an object—either matter or a ray of light—from entering the event horizon from outside, but once it has done so, gravity will prevent it from passing through this surface in the other direction and reaching the outside universe again. Loosely speaking, we can say that the event horizon *is* the black hole.

The event horizon is not a physical thing with a real material existence; on the contrary, it is an imaginary mathematical surface. All the matter that creates a black hole's enormous gravity is concentrated in a region called the singularity, which lies in the center of the black hole. According to general relativity, the singularity is a mathematical point, and the matter in it has infinite density.

There are good reasons for thinking that this infinite density does not occur. If physicists possessed a usable theory of quantum gravity and applied this theory to conditions within a black hole, they would most likely discover that the singularity was "smeared out" to some extent, and that the density was very large, but not infinite. If the singularity is not the dimensionless point prescribed by general relativity, it most likely is very small. There is every reason to think that the matter within a black hole is compressed by gravity to a volume much smaller than an atomic nucleus. Furthermore, any matter that falls into a black hole should be captured by the gravity of the singularity. For example, if an astronaut somehow managed to survive a passage through the event horizon, he could only expect that his ship would fall into the singularity and be crushed out of existence.

At least this would be his fate if the black hole was not rotating. However, this particular assumption, that a black hole would have zero rotation, is not particularly realistic, since virtually all the objects in the universe have a spin of some kind. The earth rotates on its axis, as do the other planets; the sun has a rotation, like other stars; entire galaxies rotate. It would be unreasonable to think that a black hole, which is formed by the collapse of a star that was presumably also rotating, should have no spin at all.

During the 1960s, mathematical theories of black-hole structure were worked out, and physicists discovered that the singularity in a rotating black hole would not have the form of a point, but that of a ring. Furthermore, certain theoretical calculations seemed to indicate that if an object (a spaceship, for example) happened to fall toward the singularity in just the right manner, it would miss the singularity and pass into some previously unknown region of space.

In other words, there could be a wormhole that connected the interior of a black hole to another universe, or to a distant region of our own universe. For that matter, the wormhole could lead to our universe at another epoch of time. These results seemed to indicate that travel through black holes could theoretically be utilized for instantaneous, faster-than-light travel to other regions of the universe, or for travel into the past or future.

Although this idea was extensively used by science-fiction writers, it soon became apparent that it would never work in practice. In fact, there were at least a half-dozen things wrong with it. In the first place, any astronauts who ventured near the event horizon of a black hole would likely be killed by the enormous gravitational forces they encountered. This would happen before they even crossed the event horizon. Their ship would disintegrate, and gravity would pull their bodies apart.

If astronauts could somehow manage to survive a journey

into a black hole, and if they did manage to travel through the wormhole, they would only emerge in a black hole somewhere else. Even if this could be avoided—if, for some reason, they found themselves somewhere other than in a black hole—two-way travel would still be impossible. If they tried to return to their own region of the universe, they would find themselves back in the black hole they had originally entered, and not be able to get out.

Furthermore, there were theoretical difficulties with the idea of black holes as gateways to other regions of spacetime. It was possible to question the idea that the wormholes that supposedly connected black holes really existed. According to some physicists, the mathematical abstractions that led to this conclusion were questionable. Even if the wormholes did form, they would not exist long enough for astronauts to pass through them; calculations indicated that they would close off almost as soon as they were created. And if the wormholes could somehow be stabilized and kept open, they *still* could not be used for travel through space or time, because the radiation inside them would be so intense that any being that tried to pass through one would be killed almost instantly.

Finally, the whole idea seemed paradoxical. If such wormhole travel were possible, astronauts could journey into the past. And of course, this would make it possible for them to return to earth and kill themselves as babies, or kill their mothers before they were born.

SCHWARZSCHILD WORMHOLES

Wormholes connecting black holes are not the only kind that theoretically might exist. It so happens that there exist solutions to the equations of general relativity that allow for the possibility of wormholes connecting different regions of spacetime where no black holes are present. These are called

Schwarzschild wormholes after the German astronomer Karl Schwarzschild, who did some important work on general relativity during the years immediately following the publication of Einstein's theory. I should point out, however, that the first scientist to recognize that general relativity allowed the existence of such wormholes was not Schwarzschild, but the Viennese physicist Ludwig Flamm.

Now, if Einstein's equations allow the existence of Schwarzschild wormholes, it does not necessarily follow that they exist. The situation is analogous to that of the existence of tachyons, whose existence is allowed by special relativity, but which nevertheless are not found in nature. In fact, Schwarzschild wormholes almost certainly do not exist. Calculations indicate that, if they were to exist today, the universe would have had to have been created in an unlikely, almost pathological state. In particular, the early universe would have had to contain numerous singularities.

It appears then that naturally occurring Schwarzschild wormholes could not be used for interstellar travel, or for journeys into the past or future—an astronaut could hardly take a trip through something that isn't there. However, as we shall see, this does not necessarily imply that we must give up on the idea of wormhole travel just yet.

GOING FISHING IN THE QUANTUM SEA

In 1988, Morris, Thorne, and Yurtsever published a paper in the journal *Physical Review Letters* in which they suggested that a highly advanced technological civilization might find ways of fishing microscopic Schwarzschild wormholes out of quantum chaos and enlarging them to macroscopic dimensions. If this could be done, and if these wormholes could be maintained, the three authors said, then travel through them might be possible after all.

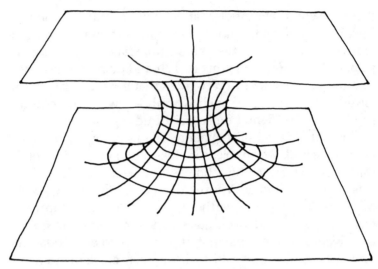

Schematic representation of a wormhole. Here, the two regions of space joined by the wormhole are represented by flat, two-dimensional sheets. If such wormholes exist in nature, they would most likely be found only at the submicroscopic level, and could not be detected by present-day scientific instruments. Morris, Thorne, and Yurtsever suggest, however, that a highly advanced civilization might be able to "fish" such wormholes out of the quantum sea and enlarge them.

Though scientists do not yet possess a workable theory of quantum gravity, they have often speculated about what a theory might conceivably tell them about the structure of spacetime on a submicroscopic level. Many of them think that, at dimensions of the order of 10^{-33} centimeters, there must be fluctuations of spacetime that are roughly analogous to the quantum fluctuations that are responsible for the creation of virtual particles.

It is likely that, at the submicroscopic level, space and time no longer have the smooth appearance that they seem to have in the macroscopic world. It is thought that, if scientists could see objects with dimensions of 10^{-33} centimeters or less, they would find that spacetime had a foamlike appearance. It would seem to be a churning mass full of tiny spacetime bridges and wormholes that would continually come into existence out of nothingness, expand, contract, twist into odd shapes, and then disappear again. Many scientists suspect that spacetime resembles a stormy, churning ocean that seems smooth only when seen from a distance.

Morris, Thorne, and Yurtsever speculate that an advanced civilization might somehow be able to "fish" a wormhole out of this quantum sea and enlarge it to macroscopic dimensions. They do not specify how this might be done; they only imply that they know of no laws of physics that would forbid it.

They then consider the question of whether such a Schwarzschild wormhole could be used for journeys in space and time. Such a wormhole would resemble the ones that Flamm and Schwarzschild described, for they would not be connected to black holes. The three physicists find that travel through such a wormhole would not be easy. Their calculations indicate that the wormhole would have to be held open by an "exotic material" or an "exotic field" capable of withstanding pressures of approximately 10^{37} pounds per square inch. Such fields, the authors admit, might turn out to be theoretically unattainable. For that matter, if such macroscopic wormholes could be created, they might turn out to be too unstable to use for travel. Alternatively, the exotic fields that held them open might interact with ordinary matter in such a way as to prevent human travel. Few people would want to make such a journey if the forces exerted by these fields were likely to turn them into something resembling strawberry jam, or so Morris and Thorne conclude in a second article, published in the *American Journal of Physics*. Finally, the backward time travel permitted by such wormholes could prevent their construction in "some as yet unimagined way."

In spite of all this, the authors maintain that the idea of traversible wormholes must be considered to be at least a possibility. "We do not know today," Morris and Thorne say, "enough to either affirm or refute these difficulties, and we correspondingly cannot rule out traversible spacetime wormholes."

Wormholes that could be used to travel in space and time sound more like science fiction than science. In a way, this is exactly what they are—in fact, the theory made its first appearance, not in a scientific journal, but in a science-fiction novel. It seems that, when author Carl Sagan was writing his novel *Contact*, he asked one of the physicists if the latter could provide a plausible method for faster-than-light interstellar travel. Sagan was provided with some of the details of the theory, which he duly incorporated into his book.

Although there are certain science-fiction aspects to the theory, we should not assume that there is no scientific purpose in pursuing such ideas. To be sure, the three authors of the theory were not trying to discover any previously unknown truths, but they were concerned with trying to establish what kinds of phenomena the known laws of physics might allow. In that sense, the questions they raise have some real interest.

11

Physics and Metaphysics

IF SCIENTISTS did not make certain philosophical assumptions, assumptions that are not susceptible of proof, it would be impossible for them to make any sense of the phenomena they observe in the natural world. It would be impossible to do physics, for example, if they did not assume that there were such things as physical laws, and that these laws always remained the same. The scientific way of thinking is so familiar to us today that we tend not to realize that it is not so obvious that this must be the case. Nature is full of variable quantities: the sun does not rise at the same time every day, and ocean tides on succeeding days do not occur at the same time or reach the same height. It would seem natural to attribute such phenomena to varying causes. The idea that they can be attributed to the workings of a never-changing law of gravitation is really rather subtle and sophisticated.

Other philosophical assumptions must be made by the physicists and cosmologists who attempt to understand the properties of the universe. For example, there is no way to demonstrate that the laws of physics must be precisely the same in distant galaxies as they are in our own region of space, but if we did not assume this, there could be no such thing as astrophysics.

Similarly, if we are to speak about the past evolution of the universe, we must assume that the physical laws operating today are the same ones that determined the behavior of fields and particles billions of years ago. Again, there is no way demonstrate that this must be the case. It is conceivable, that there never was any big bang or inflationary expansion, that we have been deceived into thinking that these events took place because we don't know that the laws of nature have changed over the course of time. The idea is conceivable, but not very appealing. If laws changed in unknown ways, we could hardly speak of "laws of nature" at all.

Though the idea that the laws of nature that we perceive are the same laws that operate in other places and in other times cannot be proved, there is a great deal of circumstantial evidence in its favor. Making this assumption has led to the creation of theories that have predictive power, and which seem to give consistent explanations of the phenomena we observe in the universe today. Making this basically philosophical assumption, in other words, has led to the creation of scientific theories that seem to make sense.

Perhaps this is the justification, also, for the philosophical assumption of simplicity. Since at least the time of Newton, it has generally been realized that, when we are confronted with a number of different possible explanations for a phenomenon, the simplest one almost always turns out to be correct. For example, the Copernican, or sun-centered theory of the solar system was shown centuries ago to be superior to the earth-centered, or Ptolemaic theory. If scientists assumed, as the Alexandrian astronomer Ptolemy had, that the sun revolved around a motionless earth, the motions of the other planets became much too complicated. It was far simpler to put the sun at the center of things.

Similarly, the idea that the ninety-odd chemical elements were too numerous to be the basic constituents of matter spurred physicists toward the discovery of the atom's constituents, the proton, the neutron, and the electron. The feeling that the subatomic world could not be made up of hundreds of different kinds of "elementary" particles led to the discovery of quarks, and ultimately to a search for a unified theory of the forces. We cannot really demonstrate that the fundamental structures of nature have to be simple. However, making that assumption certainly seems to work.

I should point out, however, that a "simple" theory can be mathematically very complicated. For example, the general theory of relativity, which is based on a small number of simple postulates, can produce some equations so complex that they have never been solved in the more than seventy years since the theory was propounded. Similarly, superstring theory, which, as we have seen, is based on the relatively simple idea that elementary particles are basically vibrating loops, has produced equations that physicists despair of being able to solve at any time in the foreseeable future.

If simple postulates sometimes lead to formidable mathematical difficulties, complicated assumptions would lead to theories that were even worse. Most likely, it would be impossible to make any sense of them at all. It would not be inaccurate to say that, without the assumption of simplicity, no one could do physics.

THE ANTHROPIC PRINCIPLE

I don't propose to discuss the assumptions of simplicity and of the constancy of the laws of nature in any great detail.

This topic has been considered by numerous other authors, and I doubt that I would have very much to add. My only purpose in mentioning them is to make the point that philosophical ideas do play a role in scientific thought, and are indeed sometimes incorporated into scientific arguments. For example, as we have seen, physicists discuss the possible violation of causality when discussing the possibility of travel backward in time, and often cite violations of causality as a reason for thinking that tachyons, for example, do not exist. The concept of causality is a philosophical idea, not a scientific one—we could include it among the philosophical assumptions upon which science is based.

In recent years, there is another philosophical idea that physicists have appealed to on numerous occasions, and which has become somewhat notorious as a result. Known as the anthropic principle, it has been denigrated by philosophers as a symptom of a return to Pre-Copernican thinking, and criticized by physicists as unscientific. The late Heinz Pagels, for example, characterized is as "a sham . . . that has nothing to do with empirical science," and called it an example of "cosmic narcissism," but other scientists have felt that the principle had real explanatory power. In fact, in their book *The Anthropic Cosmological Principle*, the British astronomer John D. Barrow and the American physicist Frank J. Tipler attempt to show that anthropic arguments have been successfully used throughout scientific history.

WE EXIST!

One of the most obvious and indisputable facts about the universe is that it contains intelligent observers. The universe may or may not harbor many different forms of intelligent life. Though many scientists suspect that life probably exists in numerous different star systems, there is no way to demonstrate this. However, one fact is clear: The universe contains at least one intelligent life form, the human species.

Yet it appears that the existence of a universe capable of harboring life is very improbable indeed. Presumably there is no reason why the laws of physics and the constants of nature could not be slightly different from what they are. For example, gravity could be a little stronger than it is, or the strong or weak forces could presumably be a little weaker. There appears to be no fundamental reason why the mass or charge of the electron should not be a little greater than it is, or the mass of the proton a little less. However, if any of these changes were made, a lifeless universe would almost certainly be the result. It appears that, unless there is some mysterious principle at work, life is the result of a series of remarkable coincidences.

The very existence of the elements upon which life is based, such as carbon and oxygen, seems to depend upon what can only be described as an unexpected stroke of luck. These elements could never have been created in significant quantities if the nuclei of carbon and beryllium atoms did not—apparently by chance—possess exactly the right energy levels. A carbon nucleus, which consists of six protons and six neutrons, can be synthesized from three helium nuclei (which have two protons and two neutrons each). However, this process would not occur very often if there did not exist an unstable form of beryllium (with four protons and four neutrons) that possessed exactly the right properties.

According to quantum mechanics, an atom or an atomic nucleus cannot possess any arbitrary quantity of energy. Both atoms and nuclei have numerous different energy levels. They cannot absorb or emit energy in any arbitrary amount, but must undergo quantum jumps that take them from one allowed energy level to another. Energy levels play quite an important role in nuclear reactions. In particular two helium nuclei would come together to form a beryllium nucleus very infrequently if beryllium did not have precisely the right energy level. The existence of this level, in effect, gives the two helium nuclei an affinity for one another that they otherwise would not possess.

The carbon nucleus, too, seems to have just the right energy level needed to enhance the formation of carbon nuclei from helium and beryllium. If this level did not exist, carbon could still be formed, but not in significant quantities. And if there was much less carbon in the universe than there is, there could not be a great deal of oxygen either. Oxygen, which consists of eight protons and eight neutrons, is synthesized when carbon and helium nuclei come together.

At this point, a skeptic might object that such arguments say nothing about the probability of life, but only demonstrate how chauvinistic human beings are. "Why," such a skeptic might ask, "should we assume that living beings must be made out of carbon and oxygen? Surely the existence of other kinds of life is conceivable."

Anyone who argued in this manner would have a point. We cannot be sure that, if life exists elsewhere in the universe, it must necessarily resemble us. For all we know, life of some kind could exist on the surfaces of red giant stars. However, the analysis of the creation of carbon and oxygen is only the first step in the argument that purports to demonstrate the improbability of life. After showing that carbon-based life is improbable, it is easy to go on to develop arguments that would seem to demonstrate the improbability of life of any kind.

It is probably reasonable to assume that life (of any kind) depends upon the existence of stars. Without stars, there would be no light or heat, and most likely no significant flow of energy from one place to another. Moreover, life almost certainly depends upon the existence of atoms, and of elements other than hydrogen. It would not be easy to conceive of living beings in a universe that contained nothing but hydrogen gas, or one in which protons, neutrons, and electrons did not frequently combine into matter.

However, in most of the universes that it is possible to conceive, neither stars nor atoms exist. For example, in our universe, the masses of the neutron and proton are very close to one another, with the neutron about 0.1 percent heavier. If this mass difference were only slightly smaller, neutrons could not have decayed into protons during the early stages of the big bang. As a result, we would have an entirely different kind of universe, one in which the relative numbers of neutrons and protons were quite different. If the proton were slightly the heavier, then protons could decay into neutrons and positrons. As a result, there would be few or no protons in the universe today. There would probably not be many electrons either, for electrons and positrons would undergo mutual annihilation when they encountered each other. In such a universe, space would be filled with neutrons and little else.

If slight changes were made in the strengths of any of the four forces, the consequences would be nearly as disastrous. For example, if the strong force were only 5 percent weaker, deuterium would not exist, since the strong force would not be strong enough to hold a neutron and a proton together. The formation of deuterium is one step in the chain of reactions by which stars convert hydrogen into helium. If deuterium could not be formed, the stars could not shine.

If the strong force were a few percent stronger, the consequences would be even worse (at least from our point of view). In this case, it would be possible to create particles called di-protons, consisting of two protons bound together. In our universe, di-protons do not exist because the strong force is not quite strong enough to overcome the electrical repulsion between the two positively charged protons. If di-protons did exist, the stars would not burn hydrogen in a slow and steady manner, as they do in our universe. On the contrary, concentrations of hydrogen gas would lead to catastrophic nuclear explosions, and stars would be blown apart before they could form. Furthermore, since hydrogen could undergo nuclear reactions so readily, the universe would consist almost entirely of helium today.

If there were slight differences in the strengths of the weak, electromagnetic, or gravitational forces, the result would also be a universe inhospitable to life. In some possible universes, there are no atoms. In others, space is filled with neutrons, or with nothing but hydrogen or helium gas. In still other universes, stars do not form, or they burn out so quickly that life never has a chance to evolve. It is even possible to conceive of universes that are inhospitable to life because they have the wrong dimensionality. For example, if space had only two dimensions, the creation of life would be difficult, to say the least. For example, an animal could not have a digestive tract that ran from one end to the other; such a passage would cut it in two pieces. If space had four dimensions, stable planetary orbits would not be possible. It can be demonstrated mathematically that, if planets did form in such a space, they would spiral into the sun.

STRONG AND WEAK ANTHROPIC PRINCIPLES

What, precisely, are we to make of the fact that we live in so improbable a universe? The anthropic principle represents an attempt to deal with that question. It can be stated in two forms. The weak anthropic principle has been stated by the British physicist Brandon Carter as follows: "What we can expect to observe must be restricted by the conditions necessary for our presence as observers." In other words, if the universe did not have the properties it does, we would not be here to see it.

The weak principle sounds like a tautology. After all, if the universe did not have these properties, there would be no one here to discuss the matter (or to propound weak anthropic principles). However, the statement is not quite as empty of content as a skeptic might think. It seems to provide an explanation for the fact that we exist at an epoch when the universe is something on the order of ten or fifteen billion years old.

A certain amount of time is required for the evolution of life. The first step is the synthesis of the elements upon which life is based, which, for the most part, are not created in the big bang. Only small traces of elements heavier than helium are formed in the processes that take place in the early universe. These elements can only be created in significant quantities in the nuclear reactions that take place within the interiors of stars.

The first stars that were formed in the universe contained little but hydrogen and helium. Over hundreds of millions of years, heavier elements including carbon, oxygen, nitrogen, and others necessary for life, were created in the interiors of massive stars. In time, these stars exploded as supernovas, and these elements were scattered through space. Eventually, a second generation of stars formed, and these elements were incorporated into the new stars, and into the planets that formed around them.

Such a process must have taken billions of years, but that was only the beginning. Before life could be created, the newly formed planets had to cool. When life finally did come into existence, it had not yet begun to evolve. It is difficult to imagine that intelligent observers could exist in a universe that was much younger than ours.

Admittedly, the weak anthropic principle can be used as a basis for arguments of this sort. Nevertheless it does seem to have a curious character, unlike virtually all the other scientific principles one encounters. According to Heinz Pagels, this difference was quite striking. Pagels argued that the weak anthropic principle was not a scientific principle at all, on the grounds that it could not be falsified.

When Pagels made this criticism, he was making reference to an influential idea that had been propounded in 1934 by the Austrian-British philosopher of science Karl Popper. According to Popper, a scientific hypothesis must be falsifiable. What makes it scientific is the fact that it can conceivably be proved wrong. For example, the statement "God exists" may or may not be true, but true or not, it is not a scientific hypothesis because it is not susceptible to disproof. Einstein's general theory of relativity, on the other hand, is scientific because it makes predictions that can be tested by experiment. If the predictions of a theory are contradicted by experiment, then the theory can be said to have been falsified.

Note that Popper's test has nothing to do with whether or not a hypothesis is correct. It is simply a definition of what is, or is not, "scientific." Pagels's point was that, if we accept this definition, then we have to conclude that the weak anthropic principle is not "scientific." It doesn't seem possible to imagine any way of disproving it.

I tend to think that we would have to agree with Pagels on this point. As a result, it isn't easy to decide what to make of the weak principle. It may be, as some physicists claim, a useful concept, or it may be, as critics like Pagels would say, only a principle that tells us what we already know (that is, that we exist), an idea that is nothing more than a kind of "cosmic narcissism." If that is all there is to the principle, how can it possibly be used to explain such things as why the universe should have the approximate age it does?

If the weak principle can be problematical, then the strong anthropic principle is even worse. The strong principle has been stated by Carter as follows: "The Universe must be such as to admit the creation of observers within it at some state." In other words, a universe that does not have the potential for the creation of life is impossible.

We may or may not agree with Pagels about the non-scientific nature of the weak principle. In the case of the strong principle, however, it would appear impossible even to argue about the matter. The strong anthropic principle obviously has too many metaphysical or theological implications; it could not possibly be considered "scientific."

If we ask *why* only universes with potential for the creation of observers can come into existence, it would seem that there are only two possible answers. Either the universe was designed by a Creator to be hospitable to life, or the observers that the universe evolves are somehow responsible for having brought it into existence.

The first possibility, incidentally, seems to be a form of the so-called argument from design for the existence of God. This argument, once quite popular, was based on the idea that the existence of God was revealed by the wonders of the natural world. Although the argument is still occasionally encountered today, it is no longer widely used by theologians. Many scholars consider that it was demolished by the eighteenth-century German philosopher Immanuel Kant.

The second possibility, that conscious observers are somehow involved with bringing the universe into existence, seems somewhat reminiscent of the philosophical doctrine of idealism, which holds that the basic substance of the universe is mind, and that the physical world is fundamentally less real. There are many different varieties of idealism, and my definition probably doesn't do justice to any of them. However, I think that it is apparent that this interpretation of the strong principle would have to be considered a kind of superidealism. After all, it would imply, not only that consciousness was more real than material reality, but also that it had played a role in creating the latter.

There may also be a third possible way to interpret the strong principle. If we were to restate it to say that the *cosmos* must be such as to admit the creation of observers, then we could interpret it to mean that there must be an infinite number of universes, some of which are hospitable to life. However, I must say that I dislike this possible interpretation more than the other two. It takes the most likely unscientific (because unfalsifiable) assumption that there are other universes, and puts it in metaphysical dress. The idea that God designed the universe may not be scientific either, but at least it is a reasonably straightforward idea that each of us can choose to believe, or not believe, according to our philosophical prejudices.

PHYSICS AND METAPHYSICS

Many scientists have the reputation of disdaining philosophy. This reputation is sometimes well-deserved. The great British experimental physicist Ernest Rutherford, for example, once commented that he considered the philosophy of his day to be a lot of "hot air." The implication, of course, was that Rutherford wasn't sitting in an armchair thinking about what the world was like, but was performing the experiments that would determine what characteristics it really had.

Rutherford's attitude may have been chauvinistic, but to some extent, it was justified. Rutherford did his most important work early in the twentieth century, at a time when philosophers were giving up on the idea of propounding all-encompassing systems, and were turning their attention to such matters as the logic of propositions, and the logic of empirical science.

On the other hand, the physicists of Rutherford's day were attaining a great deal of new knowledge. Furthermore, it was knowledge that seemed to have fairly straightforward implications. When Rutherford discovered the nucleus of the atom, for example, it wasn't necessary to puzzle about what this meant. Experiment had demonstrated that the positive charge of an atom was concentrated in a very small region, called a "nucleus," in the atom's center.

Today matters are very different. It no longer seems to be possible to do research on the frontiers of physics without confronting questions once thought to be metaphysical. Physicists have found themselves asking such questions as, Is it meaningful to speak of a time before the universe began? Did the universe have a beginning? If it did, was there any such thing as "before"? Or did time come into existence with the universe itself? What exactly is the logical status of "other universes" if these universes cannot be observed? Can we then say that they really "exist"? If we can never see the wormholes that connect our universe to the others, can we really speak, as Hawking does, of particles that attain mass by passing through these wormholes? Is it meaningful to speak of what cannot be observed? Or is this only a kind of pseudoexplanation? For that matter, what meaning should we attach to the existence of extra dimensions of space that are compacted to dimensions so tiny that they can never be observed? And, if as some superstring theorists suggest, they are not really dimensions, what are we to make of that? Finally, if an untested, possibly untestable, theory contains mathematical variables that no one can interpret, just what *does* this say about our conception of physical reality?

These are all puzzling questions. However, there is another, possibly even more significant question that looms over them, one that has to do with the growing rift between theory and experiment.

In the field of cosmology, we have a very plausible inflationary universe paradigm, which seems to explain all the major features of the universe. Yet it yields few testable predictions. If we apply Popper's criterion of falsifiability, it barely seems "scientific."

In the field of particle physics, we encounter an even more extreme situation. Many of the best theoretical physicists in the world have begun to preoccupy themselves almost exclusively with superstring theory, which has never yielded a single testable prediction, and which does not seem capable of doing that at any time in the foreseeable future. Can we really call this "science" at all? Or was Glashow right when he suggested that it was an activity akin to medieval theology?

At one time, Rutherford could call philosophy "hot air." I wonder what he would think of the present situation in physics if he were alive. Today, the boundaries between physics and metaphysics have become blurred. Questions that would have been considered metaphysical in another age enter into discussions of the origin of the universe, and physicists speak of the anthropic principles, which sometimes seem to be more philosophical than scientific. Meanwhile, some all-embracing theories are proposed that yield unverifiable conclusions, and appear similar to the metaphysical systems constantly proposed by nineteenth-century philosophers.

Some prominent physicists have found this situation alarming. They have felt that physics was drifting away from its experimental base, and was being transformed into something other than science. We have seen, for example, that some of the critics of superstring theory have been quite vehement in their denunciations of theoretical scientists who endlessly pursue untestable ideas.

THE FUTURE OF PHYSICS

I would not presume to proclaim that the members of one camp or the other are right. Perhaps it is inevitable that physics should become less of an empirical science in our day, and that speculation should take on a sometimes metaphysical character. There are limits on the experiments that it is possible to perform: for example, there are practical limits on the size of the accelerators that can be built, and naturally no government is going to provide unlimited funds for their construction. At some point, attempts to probe deeper and deeper into the structure of matter through experiment will have to come to a halt, and when experimentation reaches its limits, only theory can progress to points beyond them.

At one time, the study of the natural world was part of philosophy. We find discussions of cosmological questions in the dialogues of Plato, and there are numerous analyses of natural phenomena in the books of Aristotle. When modern science began in the sixteenth century, it appropriated questions that had previously lain in the province of philosophy. Thus it is not surprising to read the books of Galileo and discover that it was apparently not enough for him to present his theories and the results of his experiments. He constantly had to do battle with the adherents of Aris totelian doctrines as well.

As the centuries passed, the subject matter of science progressively became larger while that of philosophy contracted. By the latter half of the twentieth century, scientists were asking numerous questions that had once been considered wholly metaphysical. For example, physicists have asked such questions as, What is time? Where did the universe come from? Is creation out of nothing possible? Some even invoked the anthropic principles in an attempt to ask what, if anything, could be deduced from the fact of our existence. Physics was not the only field in which such developments took place. In the field of cognitive science, to cite just one other area of research, scientists were asking such questions as, What is mind? and, What is free will? While they speculated about the nature of consciousness and asked whether artificial intelligences could be created.

Perhaps this kind of development was inevitable. There seem to be certain basic questions that may or may not have answers, that human beings nevertheless insist on asking. Perhaps it isn't surprising that some of the people who work in the fields of physics and cosmology should now be trying to grapple with them.

Glossary

ABSOLUTE ZERO. The lowest possible temperature at which all molecular motion ceases. It is −273° C.

ANTHROPIC PRINCIPLE. The universe must have certain properties if intelligent beings are to exist to perceive it. The anthropic principle (which exists in two different forms) represents an attempt to deduce certain facts about the universe from the fact that we exist.

ANTIMATTER. *See* ANTIPARTICLE.

ANTIPARTICLE. For every particle, there exists an antiparticle. When a particle and an antiparticle come into contact, they mutually annihilate each other and energy results. Antimatter, which has not been observed to exist in nature, would be a kind of matter made up of antiparticles.

BABY UNIVERSE. According to certain very speculative theories, universes (including our own universe) may reproduce themselves by a kind of budding process. A baby universe would be one of these newly formed "buds."

BALLS OF WALL. *See* DOMAINS and DOMAIN BOUNDARIES.

BARYONS. Heavy particles such as neutrons and protons. Other kinds of baryons exist, but they are observed only in the laboratory.

BARYONIC MATTER. Matter that is made up of baryons (i.e., neutrons and protons), in other words, the "ordinary" matter of our everyday world.

BIG BANG. Scientists believe that the universe began in a very hot, highly compressed state. The initial, explosive expansion from this state is known as the big bang.

BIG CRUNCH. It is not known whether the expansion of the universe will ever slow down and reverse. If a state of contraction ever sets in, the universe may eventually destroy itself in a big crunch. This is the antithesis of the big bang.

BLACK HOLE. A black hole is the compressed remnant of a dead star in which gravity is so strong that nothing, not even light, can escape from it. *See also* EVENT HORIZON.

BOSON. A particle that transmits force. Examples are mesons, which transmit the force that binds neutrons and protons together in atomic nuclei; photons, which transmit the electromagnetic force; and gluons, which are responsible for the forces that bind quarks together.

BOTTOM-UP SCENARIO. A theory of galaxy formation in which galaxies form first, and large aggregates, such as clusters and superclusters of galaxies, form later. *See also* TOP-DOWN SCENARIO.

CHAOTIC INFLATION. A concept introduced by Soviet physicist Andrei Linde which combines the inflationary universe theory with the idea of self-reproducing universes. Naturally, this idea is very speculative. *See* BABY UNIVERSE and INFLATIONARY EXPANSION.

CLOSED UNIVERSE. A finite universe in which space closes in upon itself. Though finite, it has no boundaries. In this universe, the expansion of space will eventually halt and be followed by a phase of contraction. *See also* FLAT UNIVERSE and OPEN UNIVERSE.

COLD DARK MATTER. *See* DARK MATTER.

CONSERVATION OF ENERGY. According to this principle, developed during the nineteenth century, energy could be neither created nor destroyed. It could only be changed from one form to another. Einstein's famous equation $E = mc^2$ reveals a loophole in this principle; matter and energy can be transformed into one another.

COSMIC MICROWAVE BACKGROUND RADIATION. A background of microwaves constantly falls on the earth from every direction of space. This background radiation is a remnant of the radiation emitted in the big bang.

COSMIC STRING. There may exist defects in the structure of spacetime that would be roughly analogous to the flaws in a crystal such as a diamond. Such flaws could have a pointlike character, or they could be one- or two-dimensional. A cosmic string would be a one-dimensional flaw. Since such strings, if they exist, would be very massive, they could be the "seeds" around which galaxies formed. *See also* MAGNETIC MONOPOLE and DOMAIN BOUNDARIES.

COSMOLOGICAL CONSTANT. A constant that Einstein introduced into his equations of gravitation. It corresponds to a cosmic attractive or repulsive force that would pervade the entire universe. The cosmological constant should theoretically be very large. In reality it is either zero, or so close to zero that it cannot be measured. Scientists do not really understand why this would be the case.

CURVED SPACE. According to Einstein's general theory of relativity, space is curved. Obviously, space cannot be bent in the same way that a material object is. What the term means is that the geometry of space does not correspond exactly to the "flat" Euclidian geometry that we are taught in high school. In a curved space, for example, the sum of the angles of a triangle no longer exactly equals 180°.

DARK MATTER. It has been established that at least 90 percent of the mass of the universe exists in the form of nonluminous dark matter that cannot be observed through telescopes. Scientists are not yet sure what this dark matter is composed of. Two possibilities are that it is made up of light particles, such as neutrinos, or relatively heavy particles of one kind or another. The former are often referred to as hot dark matter because they would have emerged from the big bang at high velocities, while the latter are called cold dark matter because they would have been moving more slowly.

DEUTERIUM. A form of hydrogen in which the nucleus is composed of a proton and a neutron, rather than a single proton. The term *deuterium* is also, somewhat loosely, applied to deuterium nuclei that have not combined with electrons to form atoms.

DI-PROTON. A theoretical particle composed of two protons. The di-proton does not exist because the electrical repulsion between two protons is too strong. However, it could exist if the repulsion were slightly weaker or the strong force binding protons together slightly stronger.

DOMAINS and DOMAIN BOUNDARIES. Also known as a domain wall, a domain boundary is a two-dimensional flaw in spacetime. The name is a reference to the fact that such a wall would separate the universe into different domains. According to one recent theory, such walls might break up into balls of wall that could provide the "seeds" for the formation of galaxies.

ELECTRON VOLT. A unit of energy needed to push an electron across the voltage difference of one volt. Since it is too small a unit to be of much use, units such as MeV and GeV are more commonly used.

ELECTROWEAK FORCE. *See* FORCES.

ENERGY LEVEL. According to quantum mechanics, electrons in atoms possess only certain specific quantities of energy; they cannot possess anything in between. Energy levels also exist in nuclei, where the nucleons also possess certain energies. *See also* quantum jump.

EVENT HORIZON. A spherical surface of a black hole. Nothing that enters the event horizon can ever re-emerge, due to the black hole's gravity.

FERMION. A particle of matter. Electrons, protons, neutrons, and quarks are all fermions.

FLAT UNIVERSE. A universe in which the average curvature of space is zero. Also in this universe, space is infinite, and the expansion of space never stops. It is one that lies exactly on the borderline between open and closed universes. *See also* CLOSED UNIVERSE and OPEN UNIVERSE.

FLAVORS. *See* QUARKS.

FORCES. It has been established that there are four forces in the universe, the electromagnetic, gravitational, and strong and weak nuclear interactions. The weak and electromagnetic forces can be described as two different aspects of a single electroweak interaction. Scientists would like to find a theory that would combine all four forces within a single framework.

GEV. 10^9 electron volts. This would be a billion electron volts in American terminology. However, since the word *billion* has different meanings in the United States and in Europe, this quantity is abbreviated "GeV" rather than "BeV." Here, the "G" stands for "giga." *See also* ELECTRON VOLT.

GLUONS. The force particles that bind quarks together. *See also* BOSON.

GRAND UNIFIED THEORIES (GUTs). Theories that attempt to combine the electromagnetic, strong, and weak forces. They are rather speculative, and no one really knows which, if any, of them is most likely to be true.

GRAVITATIONAL LENS EFFECT. According to Einstein's general theory of relativity, a massive object such as a galaxy can bend light in such a way that multiple images of a distant object, such as a quasar, are created. This effect has been observed by astronomers.

GRAVITATIONAL RADIATION. A massive, gravitating body, such as a star, should radiate away a certain amount of energy in the form of gravitons. Although a number of experiments have been performed, gravitational radiation has not yet been detected. *See also* GRAVITON.

GRAVITON. The hypothetical particle that carries the gravitational force. Though gravitons have not yet been detected, physicists are sure that they exist.

GREAT ATTRACTOR. A huge concentration of mass located millions of light years away. It exerts a gravitational attraction that is drawing our Milky Way galaxy and everything else in our region of the universe toward it. But astronomers have not yet determined exactly how massive the Great Attractor is or how far away it might be.

HADRONS. Particles that feel the strong force. They can be subdivided into baryons and mesons. *See also* BARYONS and MESONS.

HIGGS MECHANISM. A theoretical mechanism for giving mass to particles. Without the Higgs mechanism, the theories that make up the standard model could not be made to work. But there is no other justification for its existence. *See also* STANDARD MODEL.

HIGGS PARTICLE. If there really is a Higgs mechanism in nature, it must manifest itself in the form of a Higgs field, and as a Higgs particle. The Higgs particle has not yet been seen, however, scientists soon hope to observe it through experiments performed with the superconducting supercollider. *See also* SUPERCONDUCTING SUPERCOLLIDER.

HOT DARK MATTER. *See* DARK MATTER.

INFLATIONARY EXPANSION. According to this theory, the universe underwent a state of extremely rapid expansion early in its history. The inflationary universe theory exists in several different forms. The original theory has been superseded by more refined versions, such as the new inflationary scenario and the theory of chaotic inflation. *See also* NEW INFLATIONARY SCENARIO and CHAOTIC INFLATION.

ISOSPIN. Sometimes it is useful to describe certain properties such as the difference between a neutron and a proton mathematically. In certain theories, protons and neutrons are regarded as a single particle—the nucleon—possessing differing amounts of a quantity called isospin. Isospin is not a real quantity, however; it is simply part of a useful mathematical technique.

LEPTON. A light particle. There are six leptons. These are the electron; two electronlike particles, the muon and tauon; and three different kinds of neutrino, one associated with each electronlike particle.

MAGNETIC MONOPOLE. An isolated north or south magnetic pole. Magnetic monopoles, if they exist, are unlike all other particles. In a certain sense, they wouldn't be particles at all, but rather pointlike flaws or defects in spacetime. Magnetic monopoles have not yet been observed.

MEV. A million electron volts. *See also* ELECTRON VOLT.

MESONS. The particles that bind protons and neutrons together in nuclei. A meson is composed of a quark and an antiquark. Many different kinds have been observed, but the one most commonly seen is the pi meson or pion. *See also* PION.

MICROWAVES. These are short-wavelength radio waves. *See also* COSMIC MICROWAVE BACKGROUND RADIATION.

MU MESON. An outmoded name for the muon (which is not really a meson). *See* MUON.

MUON. A lepton that has properties similar to those of the electron, but which is 207 times as heavy. Muons are not a constituent of ordinary matter; they are observed only in the laboratory.

NEUTRINOS. These are very light uncharged particles. It is not known whether their mass is exactly zero, or whether this quantity is simply too small to be measured. If neutrinos possess zero mass, then according to Einstein's special theory of relativity, they must move at the velocity of light. There are three varieties of neutrino: an electron neutrino, a muon neutrino, and a tauon neutrino. *See also* MUON and TAUON.

NEUTRINO OSCILLATION. If the mass of neutrinos is not exactly equal to zero, then it should be possible for neutrinos of one variety to be transformed into neutrinos of another variety. For example, electron neutrinos might spontaneously change into tauon neutrinos, and then change back again. Since the changes would presumably take place in both directions, this hypothetical phenomenon is known as neutrino oscillation.

NEW INFLATIONARY SCENARIO. An improved version of the original inflationary-universe theory developed specifically to avoid certain problems that the original theory encountered. Both theories view the inflationary expansion in much the same way. *See also* INFLATIONARY EXPANSION.

OPEN UNIVERSE. A universe in which the curvature of space is such that the universe does not close upon itself. Thus an open universe is infinite in extent. It differs from a closed universe in that the expansion of space never slows down to zero. *See also* CLOSED UNIVERSE and FLAT UNIVERSE.

PAULI EXCLUSION PRINCIPLE. This principle, originally stated by the Austrian physicist Wolfgang Pauli, said that no two electrons in an atom could be in the same energy state. The principle has since been extended to all fermions or particles of matter. The principle also implies that if two fermions are in the same energy state, they cannot be brought too close together.

PECULIAR MOTION. The component of the motion of a galaxy that cannot be attributed to the expansion of the universe.

PERTURBATION THEORY. Many of the mathematical equations used by scientists are too complicated to be precisely solved. Perturbation theory is a method of obtaining approximate solutions.

PHASE TRANSITION. A transition from one state of matter to another. Examples would be the melting of a block of ice or the boiling of water. In these two cases, ice (a solid) is changed into water (a liquid), and water (a liquid) is changed into steam (a gas). The quantum fields that permeate all space can theoretically also undergo phase transitions, changing spontaneously from one energy state to another. Such phase transitions might have played an important role in the evolution of the universe.

PHOTONS. These are particles of light. According to the quantum theory, light can be viewed either as waves or as streams of particles. Photons are also the particles that are responsible for the electromagnetic force. For example, it is the exchange of virtual photons that causes electrically charged particles to attract or repel one another. *See also* VIRTUAL PARTICLE.

PION. The pion, which is also sometimes called the "pi meson," is composed of a quark and an antiquark. The protons and neutrons that make up atomic nuclei constantly emit and absorb pions. It is this exchange of particles that causes them to stick together.

POSITRON. This is the antiparticle of the electron. It has the same mass as the electron, but possesses a positive, rather than negative, electrical charge. When a positron and an electron encounter one another, mutual annihilation is the result. The mass of the two particles is converted into energy, and a pair of gamma rays appears in their place.

PULSAR. A rapidly rotating, highly compressed remnant of a burned-out star that emits radio waves in a particular direction. If the beam of radio waves happens to sweep past the earth (as, for example, the beam of a searchlight might sweep past a ship), pulses of radio energy will be observed.

QUANTUM CHROMODYNAMICS (QCD). The theory that explains the behavior of quarks. According to this theory, quarks possess an attribute, known as color, which is the analogy of electrical charge. Quarks of different color exchange particles called gluons. Such exchanges give rise to an attractive force. The word *quantum* refers to the fact that QCD is based on quantum mechanics, while the *chromo* in "chromodynamics" is a reference to the role played by color charge.

QUANTUM ELECTRODYNAMICS (QED). The theory that explains the nature of the electromagnetic force. According to this theory, electrical attraction and repulsion result whenever charged particles exchange photons. *See also* PHOTON.

QUANTUM JUMP. Electrons in atoms can possess only certain specific quantities of energy. When such an electron undergoes a transition from one energy level to another, it is said to undergo a quantum jump. If it jumps from a higher state to a lower one, a photon of light will generally be emitted (the energy the electron gives up has to go somewhere). If it goes from a lower to a higher state, a photon of light will usually be absorbed. Quantum jumps are also exhibited by particles other than electrons. *See also* ENERGY LEVELS.

QUANTUM MECHANICS. A theory that explains the behavior of subatomic particles. It is one of the most successful science has ever known, and it is the basis of all modern physics.

QUARKS. The theoretical constituents of all hadrons, including protons, neutrons and mesons. It is thought that quarks can possess three different kinds of color charge, called red, green, and blue (but these charges bear no relation to actual colors). Quarks also come in six flavors: up, down, strange, charm, bottom, and top. Here, *flavor* is a technical term meaning "kind" or "variety." When one says that there are six quark flavors, this means only that there are six different kinds of quark.

QUASARS. The luminous cores of young galaxies. It is believed that the brightness of quasars can be attributed to radiation emitted by hot matter falling into supermassive black holes in the quasars' centers. *See also* BLACK HOLE.

REDSHIFT. When an object is moving away from an observer, the wavelengths of the light that it emits are lengthened. Since the wavelengths of red light are longer than those of any other part of the visual spectrum, there will be a shift toward the red.

RENORMALIZATION. Troublesome infinite quantities often arise in quantum field theories such as QED and QCD. Renormalization is a mathematical procedure for removing them. If a theory cannot be renormalized, it must be discarded as inconsistent. The lack of an adequate renormalization procedure has been a major stumbling block in the development of a quantum theory of gravity.

SHADOW MATTER. A hypothetical form of matter which would interact with ordinary matter only through gravitational force. It is not known whether or not shadow matter really exists. If it does exist, it can be neither felt nor seen, and can be detected only through its gravitational effects. For example, one could walk through a shadow-matter mountain and be unaware of it, or stand at the bottom of a shadow-matter ocean and have no trouble breathing normally.

SINGULARITY. If a quantity of matter were compressed by gravity into a mathematical point, that point of infinite density would be a singularity. Most likely, singularities do not exist in nature. It is probable that quantum effects would ensure that matter density never actually became infinite.

SPACETIME. A word used by physicists to describe the three dimensions of space and the one dimension of time. The acceptance of Einstein's theories caused the word to be used frequently because, in Einsteinian physics, space and time interact in a way that they do not in Newtonian mechanics. Nevertheless, it would be perfectly correct to speak of spacetime in the context of Newton's theories also.

SPARTICLES. *See* SUPERSYMMETRY.

STANDARD MODEL. This is a combination of two subtheories, the electroweak theory and QCD. It is currently the standard theory of subatomic interactions. As such, it is completely successful; no experimental evidence that would contradict it has ever been found. Nevertheless, many physicists are profoundly dissatisfied with the standard model; they feel that it does not explain enough. Many of the theories discussed in this book represent attempts to go beyond it. *See also* QUANTUM CHROMODYNAMICS and ELECTROWEAK FORCE.

STRANGE PARTICLE. Decades ago, a particle that required unexpectedly long periods of time to decay was referred to as "strange." Since those days, physicists have discovered that this "strangeness" is a quantity that can be described in a mathematical way, and the word has lost all associations with its ordinary-language meaning. Strange is also one of the six quark flavors. The strange quark is simply a constituent of certain particles that tend to have long lifespans.

SUPERCONDUCTING SUPERCOLLIDER (SSC). The SSC is a huge new particle accelerator that will be built during the 1990s. It will measure 53 miles in diameter.

SUPERGRAVITY. The name given to a number of theories that attempted to explain all four forces within a single framework.

However, none of the supergravity theories discovered by physicists proved to be completely successful, and scientists now consider the construction of superstring theories to be a more promising approach. *See also* SUPERSTRING THEORIES.

SUPERSTRING THEORIES. Many theoretical physicists now believe that all known particles might consist of vibrating loops in ten-dimensional spacetime known as superstrings. Some of them suspect that space and time themselves might be made up of superstring constituents in some manner or another. If such ideas ever prove successful, a revolution will have taken place in scientists' conception of the nature of reality. Critics of superstring theories frequently point out, however, that these theories have not produced a single quantitative prediction that can be tested in the laboratory. Some of them have gone so far as to compare the pursuit of superstring theory to exercises in medieval theology. *See also* THEORY OF EVERYTHING.

SUPERSYMMETRY (Susy). The idea that there may not be two different kinds of particles (fermions and bosons), but rather only one. If this idea turned out to be correct, the number of particles that existed in nature would be increased, not reduced. Specifically, supersymmetric theories predict the existence of an entire new class of particles, known as sparticles. To date, there is no experimental evidence to indicate that sparticles really exist.

TACHYON. A hypothetical particle which travels at velocities greater than that of light. The existence of such particles would not be inconsistent with relativity as long as they never slowed to sublight velocities. However, there is no evidence that tachyons really exist, and theories that predict their existence are now generally viewed with suspicion.

TAUON. A particle that resembles the electron and the muon, except that it is much more massive. The tauon, which possesses a mass nearly 3,500 times greater than that of the electron, is the heaviest lepton of all. *See also* LEPTON.

THEORY OF EVERYTHING. A theory from which all of the other laws of physics could be derived. Some physicists hope that one or

another of the superstring theories will turn out to be such a
T.O.E. It should be noted, incidentally, that the discovery of a
T.O.E. would not mean that the science of physics was finished. There would still remain much scientific work to be
done. Having a theory of everything would simply be like
knowing the rules of chess, and working out all its implications would be analogous to becoming a grandmaster. Not all
physicists believe that a theory of everything exists; many of
them feel that it will never be possible to sum up all physical
knowledge in a finite set of mathematical equations.

TOP-DOWN SCENARIO. A theory of galaxy formation that postulates
that mass concentrations the size of clusters and superclusters
form first. Individual galaxies are created only when these
begin to break apart.

VIRTUAL PARTICLES. Quantum mechanics allows particles to "pop"
into existence even when the energy required to create them
is not available. The energy debt created, however, must be
paid back, and these virtual particles soon disappear. Nevertheless, virtual particles have real physical effects. They are
responsible for all of the forces observed in nature.

WEAKLY INTERACTING MASSIVE PARTICLES (WIMPs). Cold dark matter would presumably be made up of WIMPs. The constituents of this matter must be weakly interacting because
strongly interacting particles would have been observed in the
laboratory by now (cold dark matter must be made up of
particles that have never been observed). Similarly, WIMPs
must be massive because particles that do not weigh very
much would travel at high velocities, and would thus be a
form of hot dark matter. *See also* DARK MATTER.

WORMHOLE. This is a pathway that joins two widely separated
regions of space. If other universes exist, it is also conceivable
that our universe could be connected to them by wormholes.
Wormholes are merely a theoretical concept. They have not
actually been observed. In fact, their dimensions may be so
small that it would not be possible to observe them.

Selected Bibliography

Barrow, John D. and Frank J. Tipler. *The Anthropic Cosmological Principle*. Oxford: Oxford University Press, 1988.

Carrigan, Richard A., Jr. and W. Peter Trower, eds. *Particle Physics in the Cosmos*. New York: Freeman, 1989.

Cohen, Nathan. *Gravity's Lens*. New York: Wiley, 1988.

Cooper, Necia Grant and Geoffrey B. West, eds. *Particle Physics: A Los Alamos Primer*. Cambridge: Cambridge University Press, 1988.

Davies, P. C. W. and J. Brown, eds. *Superstrings: A Theory of Everything?* Cambridge: Cambridge University Press, 1988.

Davies, Paul, ed. *The New Physics*. Cambridge: Cambridge University Press, 1989.

Davies, Paul. *Superforce*. New York: Simon & Schuster, 1984.

Feynman, Richard P. *QED: The Strange Theory of Light and Matter*. Princeton: Princeton University Press, 1983.

Gribbin, John. *In Search of the Big Bang*. New York: Bantam, 1986.

————. *The Omega Point*. New York: Bantam, 1988.

Hawking, Stephen. *A Brief History of Time*. New York: Bantam, 1988.

Kaku, Dr. Michio and Jennifer Trainer. *Beyond Einstein*. New York: Bantam, 1987.

Morris, Richard. *Dismantling the Universe*. New York: Simon & Schuster, 1983.

————. *Time's Arrows*. New York: Simon & Schuster, 1985.

Pagels, Heinz. *Perfect Symmetry*. New York: Simon & Schuster, 1985.

Peat, F. David. *Superstrings and the Search for The Theory of Everything*. Chicago: Contemporary Books, 1988.

Riordan, Michael. *The Hunting of the Quark*. New York: Simon & Schuster, 1987.

Silk, Joseph. *The Big Bang*. New York: Freeman, 1989.

Trefil, James. *The Dark Side of the Universe*. New York: Scribner, 1988.

_____. *The Moment of Creation*. New York: Scribner, 1983.

Weinberg, Steven. *The First Three Minutes*. New York: Bantam, 1984.

INDEX